COMBINATORIAL MATERIALS SCIENCE

THE WILEY BICENTENNIAL–KNOWLEDGE FOR GENERATIONS

Each generation has its unique needs and aspirations. When Charles Wiley first opened his small printing shop in lower Manhattan in 1807, it was a generation of boundless potential searching for an identity. And we were there, helping to define a new American literary tradition. Over half a century later, in the midst of the Second Industrial Revolution, it was a generation focused on building the future. Once again, we were there, supplying the critical scientific, technical, and engineering knowledge that helped frame the world. Throughout the 20th Century, and into the new millennium, nations began to reach out beyond their own borders and a new international community was born. Wiley was there, expanding its operations around the world to enable a global exchange of ideas, opinions, and know-how.

For 200 years, Wiley has been an integral part of each generation's journey, enabling the flow of information and understanding necessary to meet their needs and fulfill their aspirations. Today, bold new technologies are changing the way we live and learn. Wiley will be there, providing you the must-have knowledge you need to imagine new worlds, new possibilities, and new opportunities.

Generations come and go, but you can always count on Wiley to provide you the knowledge you need, when and where you need it!

William J. Pesce

WILLIAM J. PESCE
PRESIDENT AND CHIEF EXECUTIVE OFFICER

Peter Booth Wiley

PETER BOOTH WILEY
CHAIRMAN OF THE BOARD

COMBINATORIAL MATERIALS SCIENCE

Edited by

Balaji Narasimhan
Iowa State University

Surya K. Mallapragada
Iowa State University

Marc D. Porter
Arizona State University

BICENTENNIAL 1807 WILEY 2007 BICENTENNIAL

WILEY-INTERSCIENCE
A John Wiley & Sons, Inc., Publication

Published by John Wiley & Sons, Inc., Hoboken, New Jersey.
Published simultaneously in Canada.

Library of Congress Cataloging-in-Publication Data:

Narasimhan, Balaji, 1975–
Combinatorial materials science / Balaji Narasimhan, Surya Mallapragada, Marc D. Porter.
p. cm.
Includes index.
ISBN 978-0-471-72833-7 (cloth)
1. Materials science. 2. Combinatorial chemistry. 3. Computer science. 4. Combinatorial analysis. I. Mallapragada, Surya. II. Porter, M. D. (Marc D.) III. Title.
TA403.6.N366 2008
620.1′1—dc22

2007010269

10 9 8 7 6 5 4 3 2 1

CONTENTS

PREFACE

Breakthroughs in materials science will be key underpinnings to the energy, healthcare, transportation, and homeland security needs in the 21st century. These needs will place even more stringent demands on the performance of materials than ever before. To accomplish these daunting and complex tasks requires disruptive approaches that shatter existent barriers, and combinatorial science applied to materials design, discovery, and analysis will play an important role in this process. Research groups all over the world have started making significant progress in this regard, reflecting the merits of applying combinatorial science principles to fundamental and technological issues in the design of advanced materials. This compilation is but a sampling of these efforts, providing readers with perspectives on the overall principles of the methodology.

The first chapter begins with a critical analysis of successful combinatorial science experiments as applied to materials science. Its perspective is very important as new methods are being developed to design materials and understand their properties. The next chapter focuses on experimental design, which is the first step in planning and implementation of high throughput experiments. Chapter three summarizes high throughput synthesis of discrete polymer libraries. As polymers become more broadly applicable in areas such as biomaterials, biosensors, and smart and responsive materials, designing polymer libraries is key and this chapter links well known synthetic strategies to newer methods of polymer synthesis. The next chapter is focused on the development of high throughput screening probes based on multiplexed atomic force microscopy. This tool is becoming ubiquitous in the analysis of binding events and interaction forces and this chapter summarizes work that uses atomic force microscopy to design chip-scale platforms for the study of protein-protein and protein-DNA interactions. The useful interpretation of data generated from combinatorial experiments requires the development of proper integration strategies of informatics techniques with high throughput screening data, and is the focus of Chapter five. The next chapter deals with applying combinatorial techniques together with directed molecular evolution to discover next generation catalysts. Chapter seven is focused on biomaterials design by integrating principles from parallel synthesis, rapid screening, and computational modeling. Chapter eight summarizes recent advances in the development of high throughput methods for the characterization of polymer thin films, which are widely used in many technological applications. Chapter 9 brings the book to a close by summarizing the important outcomes of some of the advances described in the previous chapters and presents some challenges for combinatorial materials science in the next several decades.

We are pleased have the opportunity to be part of the collection of works by established leaders and emerging researchers from academia, industry, and government laboratories. Several of the authors are from cross-disciplinary centers and institutes such as the National Combinatorial Methods Center at NIST and the Institute for Combinatorial Discovery at Iowa State University. These are but two examples of the incisive teaming efforts that will increasingly dominate the landscape of this emergent research field. We look forward to the next several decades where as some of the challenges are met, research in this area will begin to address the next grand challenge in materials science —rapid, atom-by-atom (or molecular) design of materials for next generation applications.

Ames, IA BALAJI NARASIMHAN AND SURYA K. MALLAPRAGADA

Tempe, AZ MARC D. PORTER

ACKNOWLEDGMENTS

We would like to place on record our sincere thanks to Jonathan Rose of John Wiley for working diligently with us on this concept and ensuring a finished product that we are all proud of. We would also like to thank Ms. Linda Edson of the Department of Chemical and Biological Engineering at Iowa State University for her secretarial support.

CONTRIBUTORS

Eric J. Amis, Polymers Division, National Institute of Standards and Technology (NIST), Combinatorial Methods Center (NCMC), Gaithersburg, MD 20899

Karen M. Ashley, Polymer Group, Department of Chemistry, Howard University, Washington, DC 20059

Kathryn L. Beers, Polymers Division, National Institute of Standards and Technology (NIST), Gaithersburg, MD 20899

James N. Cawse, Proto Life S.R.L., Via della Liberta 12, 30175 Marghera, Venezia, Italy

Jack F. Douglas, Polymers Division, National Institute of Standards and Technology (NIST), Gaithersburg, MD 20899

Jeremy D. Driskell, Institute for Combinatorial Discovery, Ames Laboratory—USDOE, Department of Chemistry, Iowa State University, Ames, IA 50011

Michael J. Fasolka, Polymers Division, National Institute of Standards and Technology (NIST), Combinatorial Methods Center (NCMC), Gaithersburg, MD 20899

Nicole K. Harris, Department of Chemistry and Chemical Biology, Rutgers University, Piscataway, NJ 08854

Alamgir Karim, Polymers Division, National Institute of Standards and Technology (NIST), Gaithersburg, MD 20899

Jeremy R. Kenseth, Institute for Combinatorial Discovery, Ames Laboratory—USDOE, Department of Chemistry, Iowa State University, Ames, IA 50011
(Present address: CombiSep Inc., Ames, IA 50010)

Doyle D. Knight, Department of Mechanical and Aerospace Engineering, Rutgers University, Piscataway, NJ 08854

Joachim Kohn, Department of Chemistry and Chemical Biology, Rutgers University, Piscataway, NJ 08854

Karen M. Kwarta, Institute for Combinatorial Discovery, Ames Laboratory—USDOE, Department of Chemistry, Iowa State University, Ames, IA 50011

Surya K. Mallapragada, Institute for Combinatorial Discovery, Department of Chemical and Biological Engineering, Iowa State University, Ames, IA 50011

Balaji Narasimhan, Institute for Combinatorial Discovery, Department of Chemical and Biological Engineering, Iowa State University, Ames, IA 50011

John D. Neill, Virus and Prion Diseases of Livestock Unit, National Animal Disease Center, United States Department of Agriculture (USDA), Ames, IA 50010

Marc D. Porter, Department of Chemistry and Biochemistry, Center for Combinatorial Science at The Biodesign Institute, Arizona State University, Tempe, AZ 85287

D. Raghavan, Polymer Group, Department of Chemistry, Howard University, Washington, DC 20059

Krishna Rajan, Department of Materials Science and Engineering, Combinatorial Materials Science and Materials Informatics Laboratory, Iowa State University, Ames, IA 50011

Julia F. Ridpath, Virus and Prion Diseases of Livestock Unit, National Animal Disease Center, United States Department of Agriculture (USDA), Ames, IA 50010

Amit Seghal, Polymers Division, National Institute of Standards and Technology (NIST), Gaithersburg, MD 20899
(Present address: Rhodia, Inc. Cranbury Research and Technology Center, Cranbury, NJ 08512)

Changwon Suh, Department of Materials Science and Engineering, Combinatorial Materials Science and Materials Informatics Laboratory, Iowa State University, Ames, IA 50011

Brandon M. Vogel, Polymers Division, National Institute of Standards and Technology (NIST), Gaithersburg, MD 20899

William J. Welsh, The Informatics Institute, University of Medicine and Dentistry of New Jersey, Newark, NJ 07101-1709

L. Keith Woo, Institute for Combinatorial Discovery, Department of Chemistry, Iowa State University, Ames, IA 50011

CHAPTER 1

Combinatorial Materials Science: Measures of Success[1]

MICHAEL J. FASOLKA and ERIC J. AMIS

Polymers Division
National Institute of Standards and Technology (NIST)
Combinatorial Methods Center (NCMC)
Gaithersburg, Maryland

1.1. INTRODUCTION: THE MOTIVATION FOR COMBINATORIAL MATERIALS SCIENCE

Throughout its history, materials science has been accomplished with the explicit or implicit backdrop that materials can be improved for human use. To a considerable extent, this milieu has governed the systems considered by the discipline, and the kinds of knowledge that materials scientists generate. This technological undercurrent accounts for a thread common to materials science since its earliest days, which the is study of complex systems. For example, our understanding of multicomponent phase thermodynamics would arguably not be as advanced, nor as deep, as it is today without the desire to produce improved metallurgical alloys. Certainly, this technological interplay with complexity could be restated for any number of historical cases and accomplishments across the materials science spectrum—from doped semiconductors to polymer blends.

This trend continues for today's materials scientists, as we strive to understand, and to use, increasingly complex materials systems. In this respect, the discovery, development, and optimization of today's new materials are met by three interrelated challenges (Fig. 1.1):

[1] Official contribution of the National Institute of Standards and Technology; not subject to copyright in the United States.

Combinatorial Materials Science, Edited by Balaji Narasimhan, Surya K. Mallapragada, and Marc D. Porter

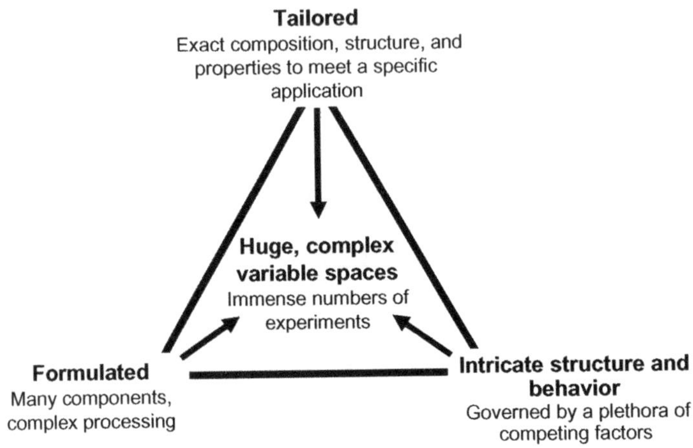

Figure 1.1. Challenges to materials development.

1. Advanced materials are often highly *tailored*, meaning that composition, structure, and properties are optimized to meet a specific application. For example, materials for fuel cell membranes [1] must transport specific ions, and they must also be structurally sound, chemically resistant, and amenable to processing. Given these requirements, it is understandable that there are only a few viable materials for fuel cell applications.
2. Today's materials are usually *formulated* from a number of components, and the structure and properties of these formulated materials can be highly sensitive to constituent levels and processing routes. Although this sensitivity makes them amenable to *tailoring*, the multivariate nature of formulations makes optimization difficult.
3. Today's materials exhibit *intricate structure and behavior*. Tailored, formulated materials often rely on structural hierarchy that includes atomic, molecular, and mesoscopic organization. Structure must often be characterized on multiple scales, and the performance of these materials can hard to measure or predict since it depends on many competing and complementary factors.

These challenges mean materials researchers are faced with large and complex variable spaces, and the reality that a huge number of experiments are needed to understand and develop materials. Materials research is expensive and time-consuming, with estimates for the time to discover and develop a new material ranging from 2 to 10 years, and with R&D costs often in excess of $20M per new material product [2].

The last 15 years (as of early 2007), or perhaps the past 40 years if we consider the earliest appearance of the concepts in the literature [3], have seen the emergence and application of so called *combinatorial materials science and high-throughput methods*. Correctly applied, these concepts have the potential to meet the challenges of developing materials [4–6]. Combinatorial materials science and high-throughput methods present means to accelerate materials research through a new experimental paradigm. Salient aspects of this new scheme are illuminated by comparison to traditional experimentation: (1) traditional experiments utilize specimens that express a single point in parameter space, but combinatorial methodology employs sample *libraries* that cover a multitude of points across parameter space (when designed effectively, combinatorial libraries explore a range of parameters in a rational and reliable manner), and (2) where traditional experiments test and analyze samples in a "one at a time" mode, combinatorial libraries are best complemented by *high-throughput measurements* that assess multiple library elements in parallel, or through a rapid serial approach.

Conceptually, it is obvious that the coalescence of these two aspects of "combi" can result in materials research that is rapid and comprehensive in its scope. This promise accounts for the intense interest in recent years surrounding these methods. Yet, it is equally obvious that the realization of these goals for materials discovery and materials science relies on how the methods are implemented. In this respect, strategies often are driven and structured by a priori visions of what *success* in combinatorial materials science entails. On this point the philosophy of combi comes into question.

1.2. THE "CLASSICAL" VISION OF SUCCESS WITH COMBI

Undoubtedly, the initial enthusiasm and inspiration for adapting combi methods to materials was rooted in comparisons with the pharmaceutical industry, which made an unreserved shift to combinatorial chemistry and high-throughput screening approaches in the early 1990s [7–9]. The primary goal of making drug *discovery* faster and more efficient strongly influenced the early vision of success and they have lasting effect on the implementation of combi today. As a noted pioneer states, "this methodology must be constructed a priori such that there are no bottlenecks, the mantra among professionals being 'screen in a day what you synthesize in that day, and analyze in a day what you screen in that day.'" [10] According to this picture, the success of the application of the method is measured largely by the number of experiments it can accomplish, the speed of these experiments, and the design for seamless operation and efficiency.

Following the trend in pharma that places a premium on speed and efficiency, this "classical" vision for combi has driven the implementation of so-called "combinatorial workflows," which are highly developed systems for automated library preparation, measurement equipment, and analysis routines, all assembled to perform according to a well-defined experimental protocol. While many workflow designs have been posited for combi, they have a common set of components and aspects. For illustration purposes, we will discuss an example scheme (Fig. 1.2), where the workflow is a cycle with steps of *library design*, *library fabrication*, *measurements,* and *analysis.* As visualized in the scheme, the workflow is tied together by an *informatics* system.

The *library design* step is most akin to traditional "design of experiment" (DOE) activities, but modified to accommodate multivariate parameter spaces. Basically, library design determines the materials properties of interest and the portion of variable space that the combinatorial library will include. Exact parameters for library array elements are defined, as well as statistical replicates and reference elements. Library design can include aspects of statistical DOE in an effort to illuminate parameter interrelationships.

Library fabrication is the process of physically producing the combinatorial array. In a well-developed workflow, library fabrication is fully automated, including coordinated operation of devices for materials handling, metering, mixing, and sampling, as well as a means for varying processing conditions over library position and/or time. In addition to accurately reflecting the library design, the fabrication route must couple with measurement and analysis steps. As we discuss below, the latter requirement can be difficult to achieve since a single library design may not be adequate if measurements necessary to characterize a library change or increase.

Measurements in combi workflows are by necessity high-throughput, since libraries can exhibit hundreds or thousands of cases. This can involve development of new instrumentation or the adaptation of existing devices for high-throughput operation. In a workflow scenario, the *measurement* step is met by additional challenges of automating and coordinating the set of instruments needed to sufficiently characterize libraries. System development must accommodate the fact that measurement instruments (whether custom-made or from a vendor) may not be designed for incorporation into a workflow.

Today, most scientific data analysis is computer-aided, if not automated. However, in the vision of the combi workflow, the *analysis* stage goes beyond performing scientific calculations and data handing in a faster manner. The analysis stage can involve data-mining schemes and multivariate statistical

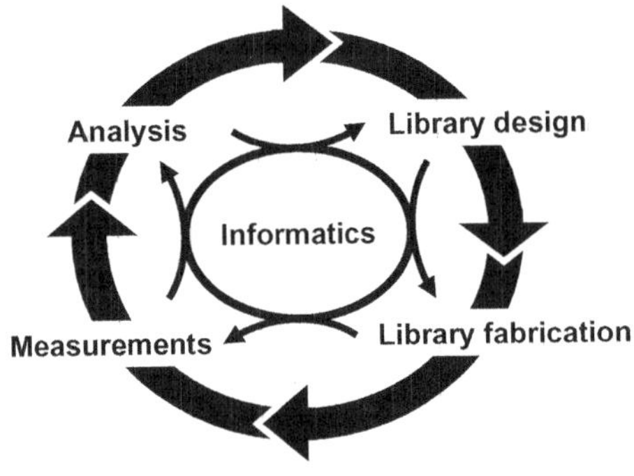

Figure 1.2. Example combinatorial workflow.

treatments for illuminating trends and correlations in the library data space. These difficult routines can be necessary if the library design does not preclude superfluous data points or if the library changes too many parameters at once. An important and challenging aspect of combi analysis is the visualization of combi datasets, which are large, complex, and can consist of a variety of data types including single values, spectra, and images. Finally, a goal of workflow analysis is to provide parameters for the design of libraries, perhaps with a refined or expanded scope, for further rounds of the combi cycle (Fig. 1.2). This "feedback" mechanism is widely termed "closing the combi loop." In the classical view of combi workflows, such feedback is one hallmark of a successful system.

A combi workflow depends on a sophisticated *informatics* infrastructure that coordinates and cements the workflow steps into a functional whole. An informatics infrastructure integrates functions such as DOE, instrument automation, data collection, automated analysis, datamining, and data visualization around a (typically central) database that is structured for research [11,12]. Constructing a combi informatics infrastructure is a formidable challenge, especially if seamless operation is desired. It can take years to develop and requires dedicated, expert personnel to achieve and maintain.

The combi workflow vision is an extremely valuable one, and there is no doubt that significant materials discoveries and knowledge generation have occurred when they are implemented properly. Nevertheless, when considering the concept of success in combinatorial materials science, and of workflows representing the principal realization of this idea, other factors must be considered. Primary among these is the growing body of combi-related materials research that has been pursued and published in recent years; a conservative search of the literature [13] yields nearly 1000 journal articles since 1990, with substantial growth in the number of publications each year (see Fig. 1.3).

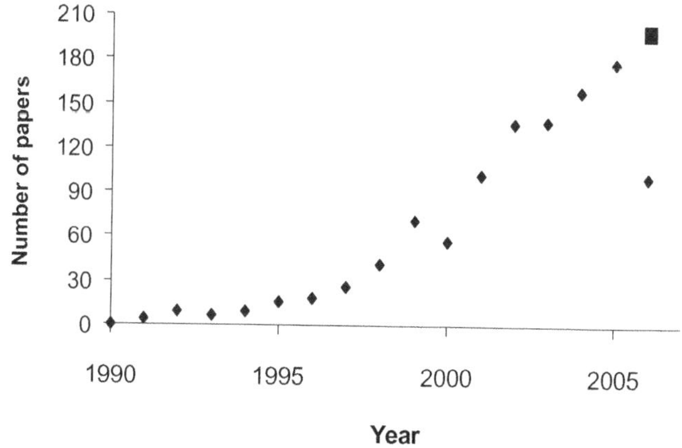

Figure 1.3. Materials combi publications versus year. 2006 value is (■) is projected based on publication count in June 2006.

However, a brief analysis of these papers reveals that only a fraction of this work results from the type of highly developed workflows discussed above. Indeed, these publications are issued from a great number and variety of institutions. These include smaller academic and industrial research groups that arguably do not have the resources to fully accomplish the workflow vision. Moreover, this literature and other reports [14] show that combi concepts are being applied to an increasingly wide set of materials systems, such as emerging technology products (e.g., nanostructured materials and organic electronics) or specialty consumer goods (e.g., personal care and cosmetics). Materials discovery in these areas is extremely fast-paced, and it is questionable whether the construction of a priori, well-developed workflows is commensurate with the rapid product turn-around required from R&D of such systems.

How do we reconcile the classical idea of combi success with these trends? One option would be to say that since they fail to match the workflow vision, most combi studies are "less than successful" or somehow incomplete. We posit that this characterization has implications that are less than positive for the field, since it suggests that (1) successful combi studies can be achieved for only a very small set of materials systems and (2) combi should be pursued only where "full" success is assured. These notions conceptually limit the scope of where combi can be useful, and amplify a sense of risk associated with pursuing combi. So, for institutions already skeptical of combi, this provides an additional excuse for "not to start," and for the initiated, it can be an excuse "not to expand." In either case, a likely result is a fewer number contributions to the state of the art in combi techniques.

Another option is to reconsider the criteria for success. A broader vision, that accommodates different styles of implementation, that focuses on the knowledge produced by combi rather than the number of specimens it can process, and that emphasizes a measured approach to infrastructure development, seems to more accurately reflect the current state of the field. More importantly, a more inclusive idea of success could reduce conceptual barriers (primarily the sense of risk) to implementing combi tools, and this could be key for sustaining the field, and for driving innovation in methods development and their application. These ideas will be discussed further below.

1.3. IMPLEMENTATION OF THE CLASSICAL VISION: WHERE HAS IT BEEN ACHIEVED AND WHY IS IT NOT MORE WIDESPREAD?

Of course, excellent workflows for have been built and some of these have been highly effective in accelerating the discovery and development of a range of materials products [5], including heterogeneous catalysts [15–18], coatings [19–22], and electronic materials [23–27]. A look at these cases (and others) reveals that for the most part well-developed workflows are built where certain conditions exist. First, we see workflows where significant human and capital resources are available specifically for combi system development or purchase. Accordingly, workflows are often a hallmark of larger companies with large R&D budgets and staffing resources, or specialty companies (e.g., Symyx, HTE, UOP) whose business is development or use of combi technology. In addition, workflows are found where materials researchers are able to leverage existing combi technologies or model processes established for other aims, such as biotechnology or pharmaceuticals. A good example is the case of combi heterogeneous catalysis research, which is widely implemented in ways analogous to pharmaceutical workflows, and which uses fundamentally similar equipment. While the extensive effort needed to modify this equipment for catalysis development should not be diminished, similar parallel chemical reactors and chemical activity sensors were already in place for several years in pharma (the pharmaceutical industry). Finally, combi workflows are found where the goals and processes of materials research are well defined and where experimentation can be accomplished with repeatable protocols. This accounts in part for the highly developed and successful workflows seen in the coatings industries, which were among the first to adapt combi for materials product R&D (for examples, see references [19,21,22,28–32] and citations therein). Indeed, due to customer demands, industrial coatings R&D can include a common set of sample preparation and test protocols, many of which can be automated as part of a combi system.

Certainly, research organizations that have implemented workflows have balanced the economic payoff of combi discovery with the time and expense required for infrastructure development. The key is to realize that in many of these cases, the scales were tipped toward workflows because the barriers to development were lowered because of ample focused resources, or because the system did not need to be "produced from scratch," due to existing technology. However, for many institutions and most materials research situations workflow-enabling conditions do not exist and persistent barriers hamper (and may even prohibit) workflow development. Primary among these challenges is library fabrication. This the major obstacle to starting combi for any specific materials set. Moreover, because of the difficulty of designing flexible library fabrication equipment, it is a problem that can reemerge each time materials research goals change. Even modest alterations in additive sets or processing routes may require extensive reengineering of equipment, plus testing of the new library fabrication process for reliability, repeatability, and so on. For example, a current trend in industrial product formulations (e.g., coatings, cosmetics, personal-care products) is to include nanostructured components (e.g., nanoparticles, nanoscale colloids, and micelles). While at first glance nanostructured components seem to be "just another additive" in a formulation, in fact they present a slew of new library fabrication chal-lenges, which have been identified by industry as key barriers to implementing combi for these systems [14]. In particular, to illuminate structure–property relationships, libraries of nanostructured formulations must be amenable to structural characterization by nanoanalysis techniques, including light and X-ray scattering, electron microscopy, and scanned probe microscopy. These methods demand highly specific sample conditions (geometry, thickness, planarity, roughness, etc.) that are not accommodated by current automated formulators. Indeed, it is likely that the rigors of nanocharacterization demand the development of entirely new library fabrication strategies.

Our example of nanostructured formulations illustrates a corollary barrier to workflow development, which is the difficulty of integrating so-called "necessary" characterization techniques into a combi system. In the case of nanostructured materials, R&D *necessarily* involves nanoscale measurements like those noted above, yet high-throughput versions of these techniques (or replacements for them) do not always exist. This point is apparent in the case of transmission electron microscopy (TEM). Because it provides essential nanoscale morphological information that cannot be achieved otherwise, TEM is relied on widely by researchers in both academia and industry. Yet, with the exception of a few advances [33,34], TEM remains incongruent with high-throughput experimentation, with no general solution in sight. Mechanical measurements, especially assessments of yield and failure, pose similar problems. In certain cases, specific tests are required because customers, or regulations, demand them. For example, in the coatings industry, "real time" aging and weathering tests are customer-trusted measures of product performance. While there has been progress in accelerated weathering testing methods [29], the current ability to predict real-time weathering properties is still limited. Accordingly, traditional measurements remain a necessity, at least on a subset of promising coatings formulations.

Informatics has been a central challenge to combi since its inception, and this remains the case today. As discussed above, workflow informatics infrastructure is complex, expensive, and time-consuming to achieve. In some respects, informatics development is faced by the same sort of barriers that inhibit library fabrication. Primarily, whether it is built in house or purchased, it is difficult to create a "general" infrastructure that accommodates changing research goals. So, in addition to expense of establishing an informatics system, extensive retooling can be required when new materials are tackled. This issue is exacerbated by several factors. First, it can be difficult to find personnel suited for combi informatics development, since it requires expertise in both computer/information science and materials science. In addition, integrating new instrumentation into an existing workflow can be difficult. Custom-built instruments require the construction of custom automation and system interoperability routines. Moreover, while most commercial instruments are supplied with control software, it is often proprietary and rarely geared for the flexible automation and system interoperability necessary for workflow integration. Accordingly, integration of commercial instruments involves building metaroutines that connect and drive vendor-supplied software. Since a workflow can contain instruments from multiple vendors, formation of a seamless informatics infrastructure is a complicated endeavor. This situation is hampered by a lack of standard data formats for interoperability, which, if they existed, could streamline infrastructure design and ease device integration. In this respect, there are some promising developments, as XML-based data formats applicable to materials combi have begun to emerge from industry [35], government [36], and academia [37,38].

1.4. AN ALTERNATE VISION OF SUCCESS WITH COMBI

With persistent barriers making the "workflow" vision unattainable in many cases, is there a more useful driving principle for the development and application of combi? What would this revised concept of success look like?

We posit that a measured, and more *immediate*, view of combi infrastructure development is a key philosophical guide. In this concept, the focus is on the effectiveness of a combi system, and its components, for knowledge generation, rather than the number of specimens it may ultimately process. As opposed to an "all or nothing" approach, infrastructure would be built with the goal of attaining benefits of combi implementation, but not necessarily a "complete," seamless workflow. This measured view of success would balance (1) the resources dedicated to *each aspect* of infrastructure development (libraries, high-throughput measurements, informatics, etc.) with the *immediate and apparent* benefits of building that piece—this "matching the hammer to the *current* nail" means that some aspects of the classical workflow model might be developed only modestly, or not at all; against (2) the time to develop infrastructure, such that it meets the R&D timescales required to effectively address the problem of interest. For example, modest infrastructure development, accomplished quickly, can result in timely combi benefits for faster-moving R&D of emerging systems. We will elaborate on these ideas next, and to make our point we will focus on the benefits that can be derived from developing individual aspects of the combi cycle.

Libraries are a solution in themselves. Combinatorial libraries can provide a convenient, compact, and powerful platform for scientific research, even if the rest of the experimental procedure is executed traditionally. When designed well, a library can be a "solution in itself" that amplifies the scientific effectiveness of sample preparation beyond the simple fact that it provides more specimens to analyze and/or measure. From a practical standpoint, since they are fabricated and processed under identical conditions, libraries can minimize errors, and solve consistency problems associated with fabricating equivalent numbers of individual specimens. Moreover, because of the smaller size of typical array elements, combinatorial libraries can minimize waste, and maximize the effective use of expensive additives, or custom synthesized components that may be in limited supply.

Most importantly, combinatorial libraries can provide scientific insight that might not be possible otherwise. By their nature, libraries allow researchers to consider "spaces" rather than individual "points," by realizing whole parameter spaces in a physical form. As an illustrative example, consider "gradient" combinatorial libraries, which systematically and continuously change in one or more properties as a function of position [23–27,31,39–58]. Gradient specimens are unique in their ability to express comprehensively an entire variable space within a single specimen, and no values are "skipped over" as with individual specimens or discrete arrays. Accordingly, gradients are unparalleled in their ability to map phase behavior, property correlations, optimum conditions and critical phenomenon—which are central goals in materials science—and do so in a single experiment. In this respect, they also can be "self-reporting," meaning that they express key results without extensive analysis. A prime example of this is seen in the gradient polymer phase diagrams developed at the National Institute of Standards and Technology (NIST), which can illuminate phase boundaries [39,40], structural changes in self-assembly [41,42], and dewetting transitions [43–45] on relatively simple visual inspection. Of course, more extensive analyses of gradient libraries yield data that are unmatched in their detail.

To be sure, a fertile library design can transform the way scientists think about specimen preparation, and conduct research. Consider, for example, the "diffusion couple" approach used in metallurgy research [59]. While "combi" was not in the nomenclature when it was conceived more than four decades ago [60], this is essentially a gradient technique, and its ease and efficacy have made it a widespread practice for generating metal phase diagrams in academia and industry. In recent years, Zhao [61,62] has pioneered the extension of the diffusion couple concept to produce ternary libraries. This approach retains the elegance of the original technique and promises similar impact.

"Cosputtering," and other codeposition approaches, are another example of an elegant, flexible, and high-impact library design route. Implementation of these "composition spread" methods requires specialized equipment. However, device designs are straightforward, and once constructed, instrumentation can produce binary and ternary (and higher-order) gradient libraries of metals, ceramics, and organics. As evidenced by a huge record in the literature, nearly any material that can be sputtered, evaporated, or coated via chemical vapor deposition is amenable to this route. For examples, see Refs. 23–27 and 63–68 and papers cited in Refs. 6 and 69. Because of this flexibility, codeposition approaches form a widespread foundation for combi in functional inorganic materials, especially in academic laboratories, where materials research targets can change rapidly. Because this library design is so fruitful, adopters are able to concentrate on other aspects of combi. As a result, academic groups who have applied these technique have produced some highly innovative combi systems [5,34,37,69–75].

A growing number of commercial measurement instruments operate in automated or high-throughput modes. These include plate readers for parallel UV-visible spectroscopy, fiberoptic probe Raman and near-IR spectroscopy, IR microscopy, atomic force microscopy, optical microscopy, nanoindentation, differential thermal analysis, and some scattering equipment. Most of these measurements are useful for materials research, but appropriate libraries are required to leverage this high-throughput instrumentation. Accordingly, a focus on library design, with the goal producing specimens that can be used with this these devices, can result in "mini workflows" that can be very productive without further system development. For example, gradient approaches often produce planar and smooth libraries that neatly complement automated scanned probe, optical and IR microscopies, and nanoindentation, as demonstrated in a number of studies [14,49,56,76–81]. A similar philosophy is seen in the clever "sector spin coating" technique, developed recently at the Dutch Polymer Institute [82], which is capable of producing polymer specimen arrays for automated atomic force microscopy (AFM) and other high-throughput film characterization techniques.

A little high throughput can go a long way. Even if they are not part of a full workflow, the development and application of automated, high-throughput measurements can have substantial benefits. Undeniably, in any research situation where measurement or analysis resources are in demand, a focus on high-throughput methods can have immediate payoff. For example, automated operation can maximize the impact of expensive, central, or "one of a kind" measurement resources since it can enable characterization of a larger number of specimens by a larger number of users. Automated measurements also offer a consistency that can reduce error, and minimize the human subjectivity associated with certain measurements, for example, the selection of areas for microscopy analysis.

Beyond these practical benefits, high-throughput measurements can have an immediate impact on scientific innovation. In many research scenarios, sample preparation is easy and fast, even without automation, and the potential rate of sample production greatly outweighs the rate of traditional analysis. In these cases, high-throughput methods can liberate researchers, as it removes the *conceptual* barrier of "which sample can I measure today?" When measurement economy is lifted, creative scientists naturally expand their idea of what is possible. Since they can test a larger set of specimens, they consider riskier, perhaps more innovative, materials cases. In this sense, even modest acceleration in measurement or analysis can be effective. From the perspective of a single researcher, the difference between 1 sample/day and 3 samples/day can be great, especially if two of those samples are being measured while he or she is thinking creatively about the next steps.

The development of high-throughput capabilities can also drive innovation in measurement science. The process of building instruments that meet the rigors of high-throughput operation (e.g., fast, automated, and flexible) can result in both fundamental improvement of existing methods, and entirely new approaches to characterize specimens. Take, for instance, the "buckling" metrology developed to rapidly measure the mechanical modulus of polymer film libraries [83]. In addition to being a quantitative high-throughput technique applicable to a wide range of polymer types, it can also serve materials that are otherwise difficult to measure. For example, in nanotechnology applications, the buckling route is highly effective for measuring nanoporous low-K dielectric films [84], and the modulus of ultrathin polymer films down to 6 nm thick [85]. More recently the technique was "reversed" to give measurements of ultrasoft systems ubiquitous in biomaterials, such as elastomers and hydrated gels [86].

A similar trend can be seen in techniques such as AFM and nanoindentation. There is no doubt that AFM instrumentation has become more robust, user-friendly, and versatile in recent years; and a case can be made that these advances were driven in large part by applications for the semiconductor industry, which demanded flexible, automated operation for higher-throughput device characterization. Similar goals may produce measurement innovations in nanoindentation. Take, for example, the ability of some commercial instruments to image the shape of the indentation site and the amount of plastically deformed material around it [87]. Ultimately, this information promises to help make nanoindentation data quantitative, and it may be the key for extending nanoindentation to viscoelastic materials like polymers. However, in the short term, this capability is being developed because it enables high-throughput *screening* that is independent of other instruments, and that is amenable to a wider range of hard materials libraries for which nanoindentation is currently suited [88,89].

Informatics—when is it necessary? As with the other aspects of combinatorial materials science, a measured amount of informatics can have immediate benefits. As discussed above with rapid measurements, the development high-throughput data analyses can provide practical advantages (e.g., error management), reduce computational barriers to creative endeavors, and provide information not otherwise possible. In these respects, the development of *image analysis* routines seems to be a particularly fruitful, perhaps necessary, aspect of materials informatics. Increasingly, the characterization of complex materials involves the analysis of complex, multifaceted images, such as multilayered micrographs (e.g., AFM), chemical micrographs (e.g., IR microscopy), scattering patterns, and multidimensional spectra. The extraction of key structural and chemical information from these rich datasets requires image analysis that is both sophisticated and scientifically sound. In addition to making it more rapid, the process of automating image analysis can fundamentally improve it along these lines. Arguably, the extra rigors of "hands-free," high-throughput operation requires a statistical robustness that might not otherwise be incorporated into a routine. This can make analysis more consistent, better able to handle weak or complex signals, and less subjective. In fact, many of the robust image analysis strategies materials researchers take for granted today were first developed because of the need for high-throughput image processing in astronomy [90,91]. For decades, astronomers have faced huge numbers of multifaceted images acquired with automated telescopes. Accordingly, they have been the historical vanguard of high-throughput image analysis. In comparison, efforts in the materials sciences are rather young and fraught with fresh challenges. For the advancement of combinatorial materials science, this provides fertile ground for motivated researchers.

The development of a wider informatics infrastructure, especially of aspects related to system integration and "feedback" mechanisms, is more problematic in terms of immediate gains, and the driving question is: "How much informatics is enough?" Of course, some informatics is necessary to realize the benefits of high-throughput measurements and library design/fabrication routes discussed above. However, as one moves beyond what is needed to support these individual elements, the development of informatics infrastructure certainly suffers from diminishing returns. As outlined in an earlier section, the integrating instruments, analysis tools, and database functions into a single system entails significant challenges that can require a great deal of effort to surmount. Moreover, the time and energy required for incremental improvements in infrastructure actually tend to increase as the system is built. This is because seamless workflow function requires close attention to the *details* of interoperability, and the number and complexity of these naturally grow as the system develops. Indeed, the benefits of informatics development for workflow systems are rarely seen until the infrastructure is complete and free of defects, and this can be a long time. Furthermore, as informatics is developed toward a seamless workflow, it naturally becomes less flexible [20]. As discussed above, workflow informatics is typically built for specific, well-defined problems and processes. When the focus of research changes, much of the hard-won infrastructure will have to be modified. Finally, while much is made of the potential of "data mining," "feedback," and "artificial intelligence" in materials informatics, these sophisticated functions are costly to develop, and at this point there are few published works demonstrating that the expense is worth it. As opposed to library fabrication and high-throughput measurements, where careful design can have immediate benefits, development of detailed informatics infrastructure is probably secondary.

1.5. CONCLUSIONS

There is no doubt that the classical "workflow" vision for combinatorial materials science is inspiring to the materials research community. As a driving goal and developmental philosophy, the workflow vision has lead to impressive methods and systems, as well as discoveries and knowledge. However, there are penalties if we consider workflows to be the exclusive measure of success in combi. The process of creating complete workflows is a daunting one, especially for researchers who are new to the field. So, the idea that combi is only truly successful if a workflow can be achieved provides many with an excuse to not begin developing or applying these exciting strategies. This decreases the flow of new ideas into the field, and ultimately, this will hamper innovation.

In contrast, we have proposed and discussed a measured view of success that balances the effort of development with more immediate benefits. This more inclusive idea encourages the development of individual aspects of combi, and flexible, more quickly assembled systems that can be realistically implemented by a wider range of materials researchers. While this balanced infrastructure may not be *the* most efficient and seamless for routine and repetitive applications, all of its elements are built because they add value in themselves, and some elements are not developed at all. Shifting away from the "all or nothing" workflow model lowers the perceptual barriers to entering into combi, and makes it an option for faster-moving research situations, such as smaller materials R&D efforts that have changing targets, and emerging materials systems where rapid knowledge is imperative. Indeed, these entrepreneurial situations could provide the influx of new ideas that will be central to sustaining the innovative spirit that has driven combinatorial materials science thus far. Most generally, however, this vision fittingly paints combi as a creative research philosophy; a way of thinking, rather than simply a capital-intensive shift in the processes. As such, it provides a fresh a priori notion, useful to combi professionals and neophytes alike; substantial benefits can be reaped when you accelerate or deepen any step in materials research. The key to attaining the benefits of combinatorial and high-throughput approaches begins by thinking beyond the single-sample paradigm, to be aware of opportunities to develop these tools, and to apply them with the wisdom that has always characterized successful science.

REFERENCES

1. Prochaska, M., Jin, J., Rochefort, D., Zhuang, L., DiSalvo, F. J., Abruna, H. D., and van Dover, R. B., High throughput screening of electrocatalysts for fuel cell applications, *Rev. Sci. Instrum.* **77**(5) (2006).
2. De Lue, N., Combinatorial chemistry moves beyond pharmaceuticals, *Chem. Innov.* **31**(11):33–39 (2001).
3. Hanak, J. J., Multiple-sample-concept in materials research—synthesis, compositional analysis and testing of entire multicomponent systems, *J. Mater. Sci.* **5**:964 (1970).
4. Amis, E. J., Xiang, X. D., and Zhao, J. C., Combinatorial materials science: What's new since Edison? *MRS Bull.* **27**(4):295–297 (2002).
5. Takeuchi, I., Lauterbach, J., and Fasolka, M. J., Combinatorial materials synthesis, *Mater. Today* **8**(10):18–26 (2005).
6. Zhao, J. C., Combinatorial approaches as effective tools in the study of phase diagrams and composition-structure-property relationships, *Progress Mater. Sci.* **51**(5):557–631 (2006).
7. Adang, A. E. P. and Hermkens, P. H. H., The contribution of combinatorial chemistry to lead generation: An interim analysis, *Curr. Med. Chem.* **8**(9):985–998 (2001).

8. Posner, B. A., High-throughput screening-driven lead discovery: Meeting the challenges of finding new therapeutics, *Curr. Opin. Drug Discov. Devel.* **8**(4):487–494 (2005).

9. Tickle, I., Sharff, A., Vinkovic, M., Yon, J., and Jhoti, H., High-throughput protein crystallography and drug discovery, *Chem. Soc. Rev.* **33**(8):558–565 (2004).

10. Weinberg, H., Foreword, in *High-Throughput Analysis: A Tool for Combinatorial Materials Science*, R. A. Potyrailo E. J. Amis (eds.), Kluwer Adamemic/Plenum, New York, 2003; pp. vii–viii.

11. Harvey, M. J., Scott, D., and Coveney, P. V., An integrated instrument control and informatics system for combinatorial materials research, *J. Chem. Inform. Model.* **46**(3):1026–1033 (2006).

12. Zhang, W. H., Fasolka, M. J., Karim, A., and Amis, E. J., An informatics infrastructure for combinatorial and high-throughput materials research built on open source code, *Meas. Sci. Technol.* **16**(1):261–269 (2005).

13. Based on a search of the ISI Web of Science database, www.isiwebofknowledge.com, keywords=[(combinatorial OR high throughput) AND materials], June 15, 2006.

14. Fasolka, M. J. and Laumeier, C. E., *NISTIR 7332: NCMC-9 Combinatorial Methods for Nanostructured Materials*, 2006 National Institute of Standards and Technology, U.S. Dept. Commerce (www.nist.gov/combi).

15. Smotkin, E. S. and Diaz-Morales, R. R., New electrocatalysts by combinatorial methods, *Annu. Rev. Mater. Res.* **33**:557–579 (2003).

16. Hagemeyer, A., Jandeleit, B., Liu, Y. M., Poojary, D. M., Turner, H. W., Volpe, A. F., and Weinberg, W. H., Applications of combinatorial methods in catalysis, *Appl. Catal. A—General* **221**(1–2):23–43 (2001).

17. Senkan, S., Combinatorial heterogeneous catalysis—a new path in an old field, *Angew. Chem. Int. Ed.* **40**(2):312–329 (2001).

18. Weinberg, W. H., Jandeleit, B., Self, K., and Turner, H., Combinatorial methods in homogeneous and heterogeneous catalysis, *Curr. Opin. Solid State Mater. Sci.* **3**(1):104–110 (1998).

19. Chisholm, B. J., Potyrailo, R. A., Cawse, J. N., Shaffer, R. E., Brennan, M., and Molaison, C. A., Combinatorial chemistry methods for coating development v. The importance of understanding process capability, *Progress Org. Coat.* **47**(2):120–127 (2003).

20. Iden, R., Schrof, W., Hadeler, J., and Lehmann, S., Combinatorial materials research in the polymer industry: Speed versus flexibility, *Macromol. Rapid Commun.* **24**(1):63–72 (2003).

21. Potyrailo, R. A., Chisholm, B. J., Morris, W. G., Cawse, J. N., Flanagan, W. P., Hassib, L., Molaison, C. A., Ezbiansky, K., Medford, G., and Reitz, H., Development of combinatorial chemistry methods for coatings: High-throughput adhesion evaluation and scale-up of combinatorial leads, *J. Combin. Chem.* **5**(4):472–478 (2003).

22. Webster, D. C., Radical change in research and development: The shift from conventional methods to high throughput methods, *JCT Coat. Tech.* **2**(15):24–29 (2005).

23. Takeuchi, I., Chang, H., Gao, C., Schultz, P. G., Xiang, X. D., Sharma, R. P., Downes, M. J., and Venkatesan, T., Combinatorial synthesis and evaluation of epitaxial ferroelectric device libraries, *Appl. Phys. Lett.* **73**(7):894–896 (1998).
24. Schultz, P. G. and Xiang, X. D., Combinatorial approaches to materials science, *Curr. Opin. Solid State Mater. Sci.* **3**(2):153–158 (1998).
25. Wang, J. S., Yoo, Y., Gao, C., Takeuchi, I., Sun, X. D., Chang, H. Y., Xiang, X. D., and Schultz, P. G., Identification of a blue photoluminescent composite material from a combinatorial library, *Science* **279**(5357):1712–1714 (1998).
26. Danielson, E., Golden, J. H., McFarland, E. W., Reaves, C. M., Weinberg, W. H., and Wu, X. D., A combinatorial approach to the discovery and optimization of luminescent materials, *Nature* **389**(6654):944–948 (1997).
27. Xiang, X. D., Sun, X. D., Briceno, G., Lou, Y. L., Wang, K. A., Chang, H. Y., Wallacefreedman, W. G., Chen, S. W., and Schultz, P. G., A combinatorial approach to materials discovery, *Science* **268**(5218):1738–1740 (1995).
28. Cawse, J. N., Olson, D., Chisholm, B. J., Brennan, M., Sun, T., Flanagan, W., Akhave, J., Mehrabi, A., and Saunders, D., Combinatorial chemistry methods for coating development v: Generating a combinatorial array of uniform coatings samples, *Progress Org. Coat.* **47**(2):128–135 (2003).
29. Chin, J. W., Nguyen, T., Gu, X. H., Byrd, E., and Martin, J., Accelerated uv weathering of polymeric systems: Recent innovations and new perspectives, *JCT Coat. Tech.* **3**(2):20–26 (2006).
30. Chisholm, B., Potyrailo, R., Shaffer, R., Cawse, J., Brennan, M., and Molaison, C., Combinatorial chemistry methods for coating development iii. Development of a high throughput screening method for abrasion resistance: Correlation with conventional methods and the effects of abrasion mechanism, *Progress Org. Coat.* **47**(2):112–119 (2003).
31. Potyrailo, R. A., Olson, D. R., Medford, G., and Brennan, M. J., Development of combinatorial chemistry methods for coatings: High-throughput optimization of curing parameters of coatings libraries, *Anal. Chem.* **74**(21):5676–5680 (2002).
32. Stafslien, S. J., Bahr, J. A., Feser, J. M., Weisz, J. C., Chisholm, B. J., Ready, T. E., and Boudjouk, P., Combinatorial materials research applied to the development of new surface coatings i: A multiwell plate screening method for the high-throughput assessment of bacterial biofilm retention on surfaces, *J. Combin. Chem.* **8**(2):156–162 (2006).
33. Lefman, J., Morrison, R., Subramaniam, S., Towards high-throughput transmission electron microscopy imaging using a multi-specimen, cartridge loading device and automated data acquisition, *Biophys. J.* **88**(1):148A (2005).
34. Bendersky, L. A. and Takeuchi, I., Use of transmission electron microscopy in combinatorial studies of functional oxides, *Macromol. Rapid Comm.* **25**(6):695–703 (2004).
35. Zech, T., Sundermann, A., Fodisch, R., and Saupe, M., Using open-source software technologies and standardized data structures to build advanced applications for high-throughput experimentation environments, *Rev. Sci. Instrum.* **76**(6) (2005).
36. Kaufman, J. G. and Begley, E. F., Matml: A data interchange markup language, *Adv. Mater. Process.* **161**(11):35–36 (2003).

37. Meguro, S., Lippmaa, M., Ohnishi, T., Chikyow, T., and Koinuma, H., Xml-based data management system for combinatorial solid-state materials science, *Appl. Surf. Sci.* **252**(7):2634–2639 (2006).

38. Adams, N. and Schubert, U. S., From science to innovation and from data to knowledge: Escience in the dutch polymer institute's high-throughput experimentation cluster, *QSAR Combin. Sci.* **24**(1):58–65 (2005).

39. Meredith, J. C. and Amis, E. J., Lcst phase separation in biodegradable polymer blends: Poly(d,l-lactide) and poly(epsilon-caprolactone), *Macromol. Chem. Phys.* **201**(6):733–739 (2000).

40. Meredith, J. C., Karim, A., and Amis, E. J., High-throughput measurement of polymer blend phase behavior, *Macromolecules* **33**(16):5760–5762 (2000).

41. Smith, A. P., Douglas, J. F., Meredith, J. C., Amis, E. J., and Karim, A., Combinatorial study of surface pattern formation in thin block copolymer films, *Phys. Rev. Lett.* **8701**(1) (2001).

42. Smith, A. P., Sehgal, A., Douglas, J. F., Karim, A., and Amis, E. J., Combinatorial mapping of surface energy effects on diblock copolymer thin film ordering, *Macromol. Rapid Commun.* **24**(1):131–135 (2003).

43. Ashley, K. M., Meredith, J. C., Amis, E., Raghavan, D., and Karim, A., Combinatorial investigation of dewetting: Polystyrene thin films on gradient hydrophilic surfaces, *Polymer* **44**(3):769–772 (2003).

44. Ashley, K. M., Raghavan, D., Douglas, J. F., and Karim, A., Wetting-dewetting transition line in thin polymer films, *Langmuir* **21**(21):9518–9523 (2005).

45. Julthongpiput, D., Fasolka, M. J., Zhang, W. H., Nguyen, T., and Amis, E. J., Gradient chemical micropatterns: A reference substrate for surface nanometrology, *Nano Lett.* **5**(8):1535–1540 (2005).

46. Xu, C., Wu, T., Drain, C. M., Batteas, J. D., Fasolka, M. J., and Beers, K. L., Effect of block length on solvent response of block copolymer brushes: Combinatorial study with block copolymer brush gradients, *Macromolecules* **39**(9):3359–3364 (2006).

47. Xu, C., Wu, T., Batteas, J. D., Drain, C. M., Beers, K. L., and Fasolka, M. J., Surface-grafted block copolymer gradients: Effect of block length on solvent response, *Appl. Surf. Sci.* **252**(7):2529–2534 (2006).

48. Ludwigs, S., Schmidt, K., Stafford, C. M., Amis, E. J., Fasolka, M. J., Karim, A., Magerle, R., and Krausch, G., Combinatorial mapping of the phase behavior of abc triblock terpolymers in thin films: Experiments, *Macromolecules* **38**(5):1850–1858 (2005).

49. Eidelman, N., Raghavan, D., Forster, A. M., Amis, E. J., and Karim, A., Combinatorial approach to characterizing epoxy curing, *Macromol. Rapid Commun.* **25**(1):259–263 (2004).

50. Wang, H., Shimizu, K., Hobbie, E. K., Wang, Z. G., Meredith, J. C., Karim, A., Amis, E. J., Hsiao, B. S., Hsieh, E. T., and Han, C. C., Phase diagram of a nearly isorefractive polyolefin blend, *Macromolecules* **35**(3):1072–1078 (2002).

51. Bhat, R. R. and Genzer, J., Combinatorial study of nanoparticle dispersion in surface-grafted macromolecular gradients, *Appl. Surf. Sci.* **252**(7):2549–2554 (2006).

52. Bhat, R. R., Chaney, B. N., Rowley, J., Liebmann-Vinson, A., and Genzer, J., Tailoring cell adhesion using surface-grafted polymer gradient assemblies, *Adv. Mater.* **17**(23):2802–+ (2005).

53. Bhat, R. R., Tomlinson, M. R., and Genzer, J., Orthogonal surface-grafted polymer gradients: A versatile combinatorial platform, *J. Polym. Sci. (Pt. B—Polym. Phys.)* **43**(23):3384–3394 (2005).

54. Genzer, J., Templating surfaces with gradient assemblies, *J. Adhes.* **81**(3–4):417–435 (2005).

55. Wu, T., Efimenko, K., Vlcek, P., Subr, V., and Genzer, J., Formation and properties of anchored polymers with a gradual variation of grafting densities on flat substrates, *Macromolecules* **36**(7):2448–2453 (2003).

56. Stafford, C. M., Roskov, K. E., Epps, T. H., and Fasolka, M. J., Generating thickness gradients of thin polymer films via flow coating, *Rev. Sci. Instrum.* **77**(2) (2006).

57. Potyrailo, R. A. and Hassib, L., Analytical instrumentation infrastructure for combinatorial and high-throughput development of formulated discrete and gradient polymeric sensor materials arrays, *Rev. Sci. Instrum.* **76**(6) (2005).

58. Potyrailo, R. A., Wroczynski, R. J., Pickett, J. E., and Rubinsztajn, M., High-throughput fabrication, performance testing, and characterization of one-dimensional libraries of polymeric compositions, *Macromol. Rapid Commun.* **24**(1):124–130 (2003).

59. Kodentsov, A. A., Bastin, G. F., and van Loo, F. J. J., The diffusion couple technique in phase diagram determination, *J. Alloys Compounds* **320**(2):207–217 (2001).

60. Kirkaldy, J. S., Diffusion in multicomponent metallic systems: 3. The motion of planar phase interfaces, *Can. J. Phys.* **36**:917 (1958).

61. Zhao, J. C., The diffusion-multiple approach to designing alloys, *Annu. Rev. Mater. Res.* **35**:51–73 (2005).

62. Zhao, J. C., Reliability of the diffusion-multiple approach for phase diagram mapping, *J. Mater. Sci.* **39**(12):3913–3925 (2004).

63. Klein, J., Lehmann, C. W., Schmidt, H. W., and Maier, W. F., Combinatorial material libraries on the microgram scale with an example of hydrothermal synthesis, *Angew. Chem. Int. Ed.* **37**(24):3369–3372 (1998).

64. Orschel, M., Klein, J., Schmidt, H. W., and Maier, W. F., Detection of reaction selectivity on catalyst libraries by spatially resolved mass spectrometry, *Angew. Chem. Int. Ed.* **38**(18):2791–2794 (1999).

65. Schmitz, C., Posch, P., Thelakkat, M., and Schmidt, H. W., Efficient screening of electron transport material in multi-layer organic light emitting diodes by combinatorial methods, *Phys. Chem. Chem. Phys.* **1**(8):1777–1781 (1999).

66. Schmitz, C., Posch, P., Thelakkat, M., Schmidt, H. W., Montali, A., Feldman, K., Smith, P., and Weder, C., Polymeric light-emitting diodes based on poly(p-phenylene ethynylene), poly(triphenyldiamine), and spiroquinoxaline, *Adv. Funct. Mater.* **11**(1):41–46 (2001).

67. Schmitz, C., Schmidt, H. W., and Thelakkat, M., Lithium-quinolate complexes as emitter and interface materials in organic light-emitting diodes, *Chem. Mater.* **12**(10):3012–3019 (2000).

68. Schmitz, C., Thelakkat, M., and Schmidt, H. W., A combinatorial study of the dependence of organic led characteristics on layer thickness, *Adv. Mater.* **11**(10):821–+ (1999).

69. Takeuchi, I., van Dover, R. B., and Koinuma, H., Combinatorial synthesis and evaluation of functional inorganic materials using thin-film techniques, *MRS Bull.* **27**(4):301–308 (2002).

70. Cui, J., Chu, Y. S., Famodu, O. O., Furuya, Y., Hattrick-Simpers, J., James, R. D., Ludwig, A., Thienhaus, S., Wuttig, M., Zhang, Z. Y., and Takeuchi, I., Combinatorial search of thermoelastic shape-memory alloys with extremely small hysteresis width, *Nature Mater.* **5**(4):286–290 (2006).

71. Takeuchi, I., Famodu, O. O., Read, J. C., Aronova, M. A., Chang, K. S., Craciunescu, C., Lofland, S. E., Wuttig, M., Wellstood, F. C., Knauss, L., and Orozco, A., Identification of novel compositions of ferromagnetic shape-memory alloys using composition spreads, *Nature Mater.* **2**(3):180–184 (2003).

72. Takeuchi, I., Long, C. J., Famodu, O. O., Murakami, M., Hattrick-Simpers, J., Rubloff, G. W., Stukowski, M., and Rajan, K., Data management and visualization of x-ray diffraction spectra from thin film ternary composition spreads, *Rev. Sci. Instrum.* **76**(6) (2005).

73. Hasegawa, K., Ahmet, P., Okazaki, N., Hasegawa, T., Fujimoto, K., Watanabe, M., Chikyow, T., and Koinuma, H., Amorphous stability of hfo2 based ternary and binary composition spread oxide films as alternative gate dielectrics, *Appl. Surf. Sci.* **223**(1–3):229–232 (2004).

74. Matsumoto, Y., Murakami, A., Hasegawa, T., Fukumura, T., Kawasaki, M., Ahmet, P., Nakajima, K., Chikyow, T., and Koinuma, H., Structural control and combinatorial doping of titanium dioxide thin films by laser molecular beam epitaxy, *Appl. Surf. Sci.* **189**(3–4):344–348 (2002).

75. Ohkubo, I., Matsumoto, Y., Ohtomo, A., Ohnishi, T., Tsukazaki, A., Lippmaa, M., Koinuma, H., and Kawasaki, M., Investigation of zno/sapphire interface and formation of zno nanocrystalline by laser mbe, *Appl. Surf. Sci.* **159**, 514–519 (2000).

76. Lin-Gibson, S., Landis, F. A., and Drzal, P. L., Combinatorial investigation of the structure properties characterization of photopolymerized dimethacrylate networks, *Biomaterials* **27**(9):1711–1717 (2006).

77. Crosby, A. J., Fasolka, M. J., and Beers, K. L., High throughput craze studies in gradient thin films using ductile copper grids, *Macromolecules* **37**(26):9968–9974 (2004).

78. Schenck, P. K., Kaiser, D. L., and Davydov, A. V., High throughput characterization of the optical properties of compositionally graded combinatorial films, *Appl. Surf. Sci.* **223**(1–3):200–205 (2004).

79. Beers, K. L., Douglas, J. F., Amis, E. J., and Karim, A., Combinatorial measurements of crystallization growth rate and morphology in thin films of isotactic polystyrene, *Langmuir* **19**(9):3935–3940 (2003).

80. Meredith, J. C., Smith, A. P., Karim, A., and Amis, E. J., Combinatorial materials science for polymer thin-film dewetting, *Macromolecules* **33**(26):9747–9756 (2000).

81. Smith, A. P., Douglas, J. F., Meredith, J. C., Amis, E. J., and Karim, A., High-throughput characterization of pattern formation in symmetric diblock copolymer films, *J. Polym. Sci. (Pt. B—Polym. Phys.)* **39**(18):2141–2158 (2001).

82. de Gans, B. J., Wijnans, S., Woutes, D., and Schubert, U. S., Sector spin coating for fast preparation of polymer libraries, *J. Combin. Chem.* **7**(6):952–957 (2005).
83. Stafford, C. M., Guo, S., Harrison, C., and Chiang, M. Y. M., Combinatorial and high-throughput measurements of the modulus of thin polymer films, *Rev. Sci. Instrum.* **76**(6) (2005).
84. Stafford, C. M., Harrison, C., Beers, K. L., Karim, A., Amis, E. J., Vanlandingham, M. R., Kim, H. C., Volksen, W., Miller, R. D., and Simonyi, E. E., A buckling-based metrology for measuring the elastic moduli of polymeric thin films, *Nature Mater.* **3**(8):545–550 (2004).
85. Stafford, C. M., Vogt, B. D., Harrison, C., Julthongpiput, D., and Huang, R., Elastic moduli of ultrathin amorphous polymer films, *Macromolecules* **39**(15):5095–5099 (2006).
86. Wilder, E. A., Guo, S., Lin-Gibson, S., Fasolka, M. J., and Stafford, C. M., Measuring the modulus of soft polymer networks via a buckling-based metrology, *Macromolecules* **39**(12):4138–4143 (2006).
87. Warren, O. L., Downs, S. A., and Wyrobek, T. J., Challenges and interesting observations associated with feedback-controlled nanoindentation, *Z. Metallkunde* **95**(5):287–296 (2004).
88. Warren, O. L., Dwivedi, A., Wyrobek, T. J., Famodu, O. O., and Takeuchi, I., Investigation of machine compliance uniformity for nanoindentation screening of wafer-supported libraries, *Rev. Sci. Instrum.* **76**(6) (2005).
89. Warren, O. L. and Wyrobek, T. J., Nanomechanical property screening of combinatorial thin-film libraries by nanoindentation, *Meas. Sci. Technol.* **16**(1):100–110 (2005).
90. Bijaoui, A., Image-analysis—transfer of techniques used in astronomy to diffraction, *J. Phys.* **47**(C-5):63–67 (1986).
91. Ahern, F. J., Digital image-analysis for astronomy, *J. Roy. Astron. Soc. Can.* **78**(5):200–200 (1984).

CHAPTER 2

Experimental Design in High-Throughput Systems

JAMES N. CAWSE

Proto Life S.R.L.
Marghera, Venezia
Italy

2.1. INTRODUCTION

High-throughput and combinatorial methods for materials discovery and optimization have presented a real challenge for the effective planning of experiments. When experiments can be run in parallel by the dozens or hundreds, the classic experimental designs for data-sparse systems must be rethought for data-rich ones. This area has been covered by several brief reviews [1,2] and one book [3], but the rapid expansion of the field justifies a thorough review of the state of the art.

In reviewing the field of high-throughput materials science, it is useful to distinguish three purposes of experimentation: searching; optimization, and knowledge seeking. Each purpose will tend to use a different experimental philosophy. In *searching*, the experimenter will try to cover as large an experimental space as possible, probably with many different qualitative factors. In *optimization* the qualitative factors will tend to be fixed and the study will focus on more detail of the quantitative factors. *Knowledge seeking* tends to search for mechanistic detail, so quantitative and time-varying elements come to the fore.

2.2. STRUCTURE OF EXPERIMENTAL SPACE

In planning high-throughput experiments in a complex experimental space, it is useful to first consider the makeup of that space. Farrusseng has pointed

Combinatorial Materials Science, Edited by Balaji Narasimhan, Surya K. Mallapragada, and Marc D. Porter

out that combinatorial experimentation can encompass the entire range of possible data types in a single experiment (Fig. 2.1) [4]. These data types can appear in both factors and responses (Table 2.1).

Once the relevant data types have been determined, the detailed structure of the experimental space must be considered. The structure of a combinatorial experiment can best be considered by examining its differences from a "classical" designed experiment. The classical DOEs (factorial designs, "Latin squares," "response surfaces," etc.) are very efficient designs for determining main effects and simple interactions. They are great for optimization but almost useless for discovery research. If we take the assumptions of classical designs and deny each one, we get an indication of the types of designs and

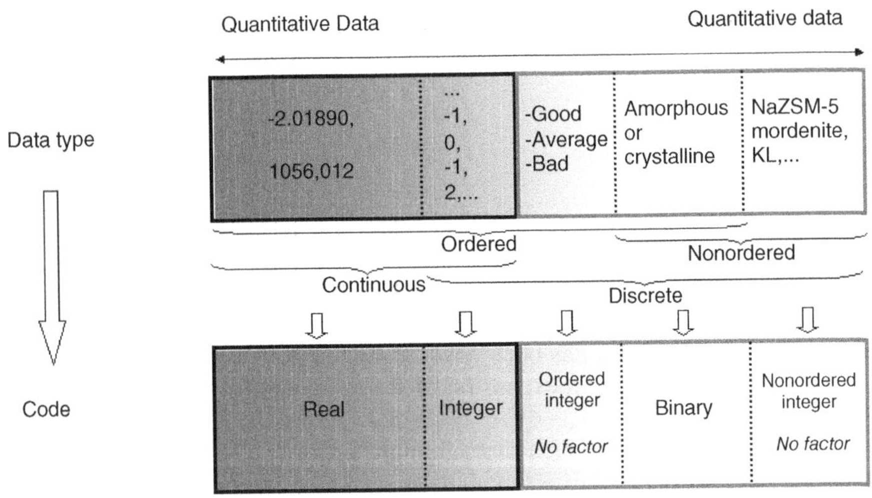

Figure 2.1. Elements of experimental space. (Reprinted, with permission, from Farrusseng, D., Baumes, L., and Mirodatos, C., in *High-Throughput Analysis*, R. A. Potyrailo and E. J. Amis (eds.); copyright 2003 © Kluwer/Academic Press.)

TABLE 2.1. Data Types Found in Combinatorial and High Throughput Experimentation

Data Types Typically Found in Factors (Independent Variables)	Data Types Typically Found in Responses (Dependent Variables)
Formulation variables	Quantitative measurements (usually real or integer)
Type of component (binary or nonordered)	Ordinal measurements (e.g., subjective scale from 1 to 5)
Amount of component (real)	Binary responses (good/bad, yes/no, lights up/doesn't)
Process variables (generally real)	
Chemical structure variables (many types)	

TABLE 2.2. Assumptions of classical DOEs and Their Systematic Violation in High-Throughput Experimentation

Classical DOE Assumptions	Violations	Resulting High-Throughput Designs
Experimental space is relatively smooth	Space isn't smooth (e.g., phase transitions)	Continuous or incremental gradients; lattices
Factors are ordered (quantitative or two-level qualitative)	Factors are not ordered (multilevel qualitative factors)	Multilevel full factorials; combinatoric methods
Simple interactions (2-way interactions and simple curvature)	⩾3-way interactions; process–formulation interactions	Multilevel full factorials; t designs; split–plot designs
Not too many dimensions (eliminate factors using screening designs)	Cannot eliminate factors early	Multilevel factorials
	All the above	Stochastic designs: genetic algorithms; Monte Carlo methods

strategies needed for high-throughput methods. A common feature of classical DOEs is that they were developed in a data-poor environment, where experiments were relatively slow and expensive [3, p 5.]. The assumptions listed in Table 2.2 were required for the efficient classical designs to work. If any of the assumptions are violated, many experiments are required. This was generally not practical until the advent of high-throughput methods.

2.3. EXTENSIONS OF CLASSICAL DESIGNS

One way in which high-throughput methods can be used is by massive extension of the classical DOE methods such as full and fractional factorial designs. These designs have historically been relatively small, typically with 16 or 32 runs to study four to eight variables. The general principles still work well with hundreds of runs and a dozen or more factors with multiple levels.

2.3.1. Factorials

Avantium Technologies is currently using very large fractional factorial and D-optimal designs as part of its standard methodology. In one report, 42 catalysts and 7 organic and inorganic bases were variables in a fractional factorial design of approximately 150–200 runs. They point out, however, that when there is a strong probability of interaction between categorical factors, any less than a full factorial design is unlikely to have the power to detect those interactions. Thus in a 12 catalyst × 4 additive × 3 solvent = 144 run full factorial experiment, the significant interactions were found, but in a reduced D-optimal design of "only" 90 runs, they were not [5]. They therefore recommend performing a "combinatorial" (full factorial) experiment in the categorical factors, a screening (fractional factorial) design on the continuous factors, and combining the insights from those two in a final optimization experiment [6].

2.4. MIXTURE SYSTEMS

Most of the experiments in high-throughput materials development involves formulation of an (hopefully) active material, be it a catalyst, phosphor, polymer blend, or magnetic material. As such, these are mixture systems and must (to some extent) follow the rules of mixture experimentation. DOEs for mixtures have been a subject of extensive statistical study, primarily by Cornell [7]. The classic design is a simplex (Fig. 2.2), covering either the entire sample space or a truncated section. These designs, as mentioned above, are effective for fitting simple linear, quadratic, or cubic models but are insufficiently dense to locate small high-value phases. Generating these dense arrays of points requires a gradient strategy.

2.4.1. Continuous-Gradient Designs

The simplest method for generating a dense gradient of points is to generate a *composition spread* composed of continuous gradients of material on a substrate (Fig. 2.3). The overlap of these gradients will form every possible combination of the ingredients. The density of the points is limited primarily by the spatial resolution of the analytical system. Since modern electronic measurement tools [8] often have a spatial resolution of a few micrometers, a composition spread of 6 cm^2 (1 $in.^2$) can contain 10^3–10^4 sampling points.

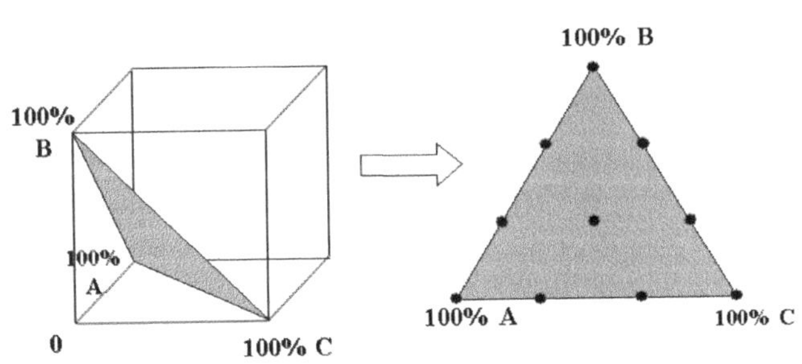

Figure 2.2. Generation of a three-component simplex from three-dimensional space.

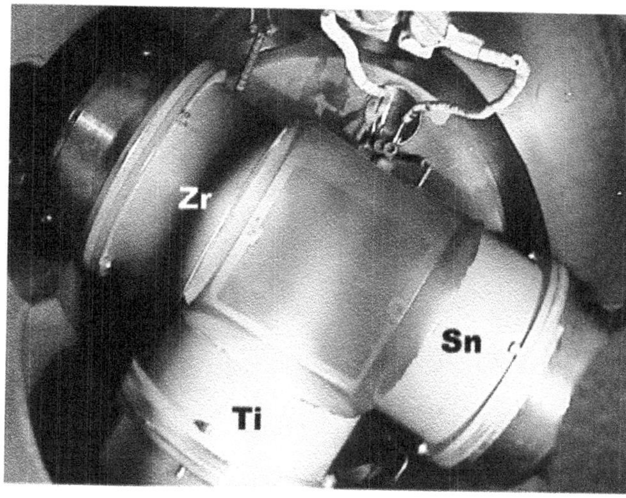

Figure 2.3. Off-axis sputtering system used for preparation of pseudoternary arrays. (Reprinted, with permission, from Schneemeyer, L. F. and Van Dover, R. B., in *Experimental Design for Combinatorial and High Throughput Materials Development*, J. N. Cawse (ed.); copyright 2002 © John Wiley & Sons, Inc.)

This method was pioneered by Kennedy in 1965 [9] and Hanak in 1970 [10], and revived by Schneemeyer and Van Dover in the 1990s [11]. It has since become one of the methods of choice for development of new electronic materials (for examples, see most of the references cited in Ref. 12). The gradients can be generated by sputtering, pulsed-laser deposition, molecular beam epitaxy, chemical vapor deposition, or simple spraying of solutions [11]. A novel method of generating these gradients using optimally shaped moving-shadow masks has been reported by Koinuma (Fig. 2.4) [13]. The key limitation of these methods is dimensional; since the substrate is two-dimensional (2D), the maximum number of active components in the system is 3.

The alternative method of generating a composition spread is by using multiple sources, masks, and moving shutters. This was pioneered and developed to a very high level by researchers at Symyx [14]. It allows a very large number of combinations of a substantial number of components, but because of its mechanical complexity, it does not appear to be as widely used as the ternary gradient method.

Another form of continuous-composition gradient is a continuously variable polymer strand formed by modulation of feeders to a microextruder (Fig. 2.5) [15]. If the feed rates are varied sinusoidally at different frequencies, the gradient will cover the design space in the form of a Lissajous figure (Fig. 2.6). The frequency ratios of the feed rates must be carefully calculated for the design to cover the experimental space effectively, especially if there are more than two feeders.

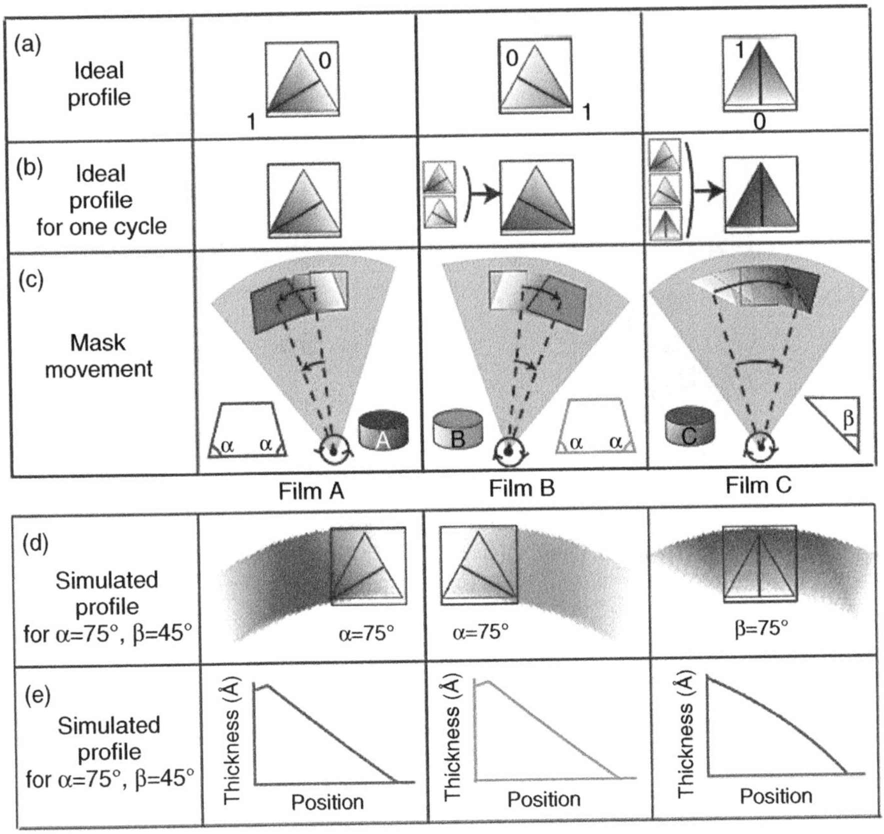

Figure 2.4. Masks for generating three-component gradient system using rotating masks. [Reprinted, with permission, from Takahashi, R. et al., *J. Chem. Inform. Comput. Sci.* **6**(1):50–53 (2004); copyright 2004 © American Chemical Society.]

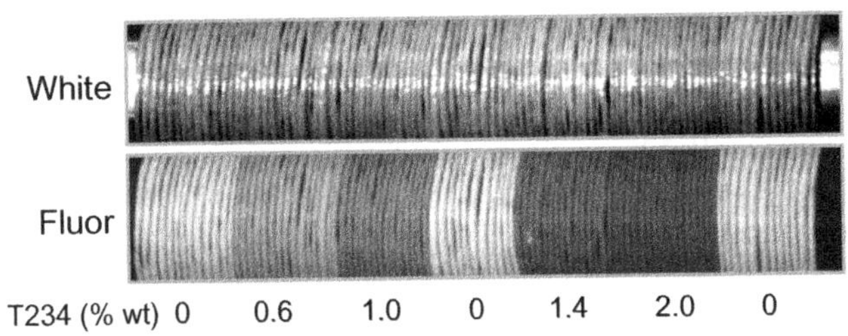

Figure 2.5. Continuously varied polymer strand. Images under white and fluorescent illumination. [Reprinted, with permission, from Potyrailo, R. A. et al., *Macromol. Rapid Commun.* **24**(1):123–130 (2003); copyright 2003 © Wiley-VCH.]

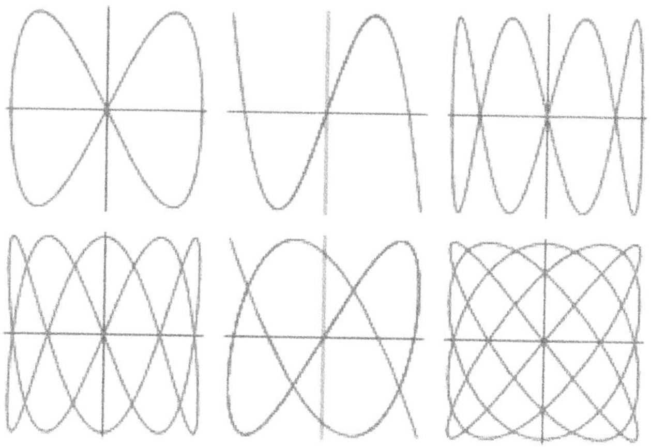

Figure 2.6. Lissajous figures potentially generated by continuously varying feed system.

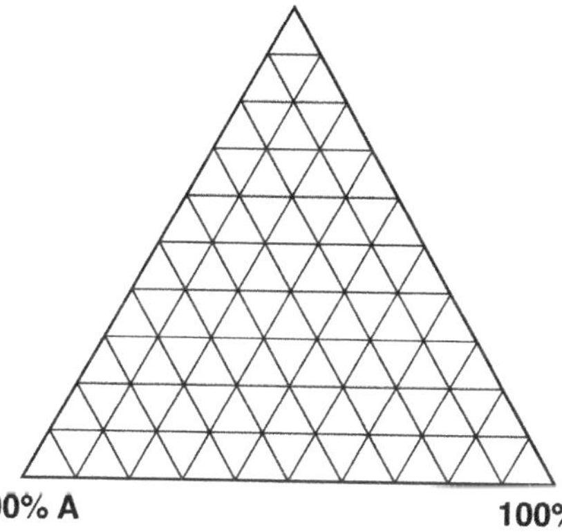

Figure 2.7. Dense (11-step) grid on a simplex. [Reprinted, with permission, from Cawse, J. N., *Acc. Chem. Res.* **34**:213–221 (2001); copyright 2001 © American Chemical Society, Washington, DC.]

2.4.2. Non-Continuous-Gradient Designs

If the number of independent formulation variables is greater than 3, or if the formulation components are not amenable to one of the projection methods mentioned above, a non-continuous-gradient method must be used to search a mixture system. This is typically a simplex arrangement with a much denser grid structure (Fig. 2.7). Each experimental point must be mixed separately, either manually, robotically or with an inkjet-type system. Mallouk has been a pioneer in this area, starting with three-factor systems [16] and building to four-factor [17] and five-factor [18] systems. His group uses an ingenious

In an important contribution to the effective use of GAs, Pereira et al. studied the effects of the variable parameters in a GA on its convergence and robustness. The test bed was a set of 1134 test reactions in a CO oxidation catalyst system. Using the response surface determined from the test reactions, they ran an 18-run designed experiment in four GA parameters (Table 2.3). They found that convergence depended strongly on population size, with a population of 8 clearly too small and results improving up to 48. Tournament selection was clearly best in one benchmark and tied with "threshold" in a second, and "elitist" selection was also clearly preferred. None of the crossover modes were clearly superior [29].

An additional parameter to be considered in a GA search for catalytic activity is the distribution of elements in the starting population. Maier [26] has suggested that

> the number of elements in a material (e.g., a single array point) should be Poisson distributed over the whole library. From Chemical knowledge it was derived that small amounts of poisoning elements lead to a small figure of merit. The chance of the inclusion of such elements increases with the number of elements in a material. This results in a decreasing chance of activity increase with an increasing number of elements.

More advanced statistical analysis of GAs is likely to yield even more understanding. Reardon [30] has shown that

> Bayesian analysis of the output of a GA suppresses the inherent stochasticity of the GA and thus makes principal components analysis (PCA) possible. PCA in turn provides a sensitivity analysis of the model parameters as well as a suitable stopping criterion for the GA.... Within the context of experimental design, this methodology easily identifies where the next experiment would be most beneficial to the development of a more robust model.

Finally, a number of authors have successfully used a hybrid strategy in which a few generations of a genetic algorithm narrows the search space to the point that the chemist can identify a "theme." This theme—a combination of elements [26,27] and a common structural feature [31,32]—is then exploited using either combinatorial or conventional methods (Fig. 2.11).

TABLE 2.3. GA Parameters Used as Factors

Elitist Selection	Population Size	Crossover Mode	Selection Method
Yes	8	1-point	Wheel
No	16	3-point	Ranking
—	24	Uniform 20%	Threshold
—	48	Uniform 50%	Tournament

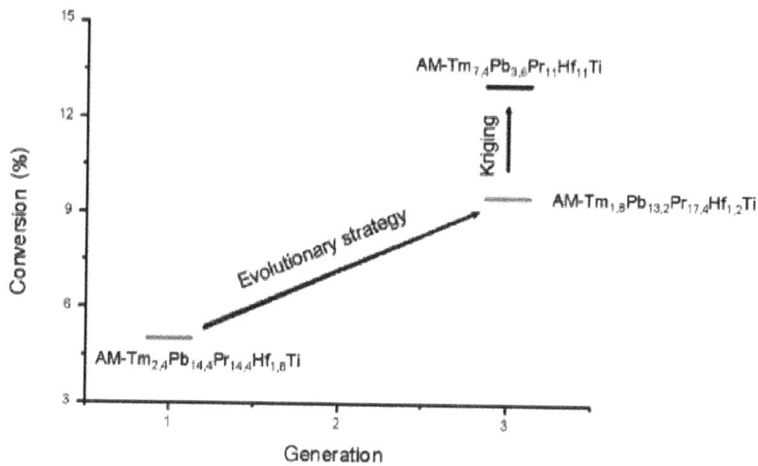

Figure 2.11. Improvement in activity of a non-noble-metal catalyst via evolutionary strategy followed by kriging. [Reprinted, with permission, from Kirsten, G. and Maier, W. F., *Appl. Surf. Sci.* **223** (2004); copyright 2004 © Elsevier B. V.]

2.6. GRID SEARCHES

In the "discovery" phase of a combinatorial chemistry program, we are typically searching for some form of activity in a "large" experimental space [1,3]. In this space, the desired activity will not be determined by a linear function of the variables. Instead, we expect to find unexpected phases or regions of high activity embedded in an uninteresting space. Search for these regions requires a thorough coverage of the entire region. The search problem may be defined as

> *Placing experimental points in such a way that the probability of one or more points being in the desired region is maximized.*

This is equivalent to minimizing the average distance from the sampling points to all possible points in the experimental space. Hamprecht [33] points out that

> It is intuitively clear that the multidimensional signal should be sampled as uniformly as possible . . . No part of the region should lie very far from the closest sample point. . . . The problem of placing points uniformly in a multidimensional space has been studied extensively in other applications and the solution is often to use a *lattice*. Which lattice to use depends on which criterion is used to measure uniformity: the *packing problem* aims at maximizing the distance between the closest pair of lattice points; the *covering problem* aims at minimizing the maximum distance between a (nonlattice) point in space and its closest lattice point.

TABLE 2.4. Preferred Lattices in Dimension 2–8[a]

Dimension	Packing Lattice	Fourier Transform of Packing Lattice	Packing Density Relative to That of Cartesian Lattice
2	A_2 (hexagonal)	$A_2^* = A_2$ (hexagonal)	1.15
3	A_3 (face-centered cubic)	A_3^* (body-centered cubic)	1.41
4	D_4	$D_4^* = D_4$	2.00
5	D_5	D_5^*	2.82
6	E_6	E_6^*	4.62
7	E_7	E_7^*	8.00
8	E_8	$E_8^* = E_8$	16.00

[a] Terminology from [Ref. 35].

The intuitively "reasonable" Cartesian lattice, with points equally spaced in all dimensions, has clearly been shown to be one of the *least* efficient lattices by all considerations [34]! Hamprecht et al. found that the preferred lattice for sampling a high-dimensional space depends on the relationship between the density of the sampling and the "roughness" of the space. If the points are so far apart that there is little or no correlation between them (sparse sampling, rough space), the best packing lattice is preferred. If the points are close enough together for them to be moderately well correlated (dense sampling, smooth space), the Fourier transform of the best packing lattice is preferred. "Sparse" and "dense" are defined only relative to the roughness of the activity function in space; a dense sampling may require more than 10^d points, where d is the dimension of the space. This would put it out of range of even the highest-throughput system for $d > 4$. The preferred lattices in dimensions 2–8 are given in Table 2.4.

The densities in the last column of Table 2.4 compare the best packing lattice (column 2) to the Cartesian. They do not relate to column 3. These packing densities illustrate how different the lattices are; but they do not prove that, for example, E_8 is better by a factor of 16. The improvement depends on the roughness of the activity function.

2.6.1. Regular Grids

Lindsey has published a series of papers in which various types of grid search are performed in an automated sequence using chemistry workstations. In his successive focused grid search (SFGS), a set of syntheses are performed on a regular grid; "the location of the optimal response obtained in one search cycle constitutes the location about which a more fine-grained search is performed" [36]. In subsequent papers the concept is expanded to sequential Simplex searches [37,38] and adaptive searches [39].

2.6.2. Kriging

The lattices presented above are defined for an infinite space. In a real space, sampling that yields the *best linear unbiased estimator* is referred to as *kriging*, after the geologist D. G. Krige. The problem of sampling a finite space must include consideration of the edge effects as well as the experimenter's best estimate of the "roughness" of the landscape. Generation of a set of sample points is a complex optimization problem. Several computer programs have been written for this. The most general is the program Gosset, from Bell Laboratories [40]; an alternative distance-based program was developed by Royle [41]. An interesting alternative can be found in the tables of "uniform designs" generated by Fang and Lin. These are relatively small designs (up to 30 samples in 2–10 dimensions) which have been tabulated on their website [42,43].

2.7. DESIGNS FOR DETERMINING KINETIC PARAMETERS

2.7.1. Optimal Sampling Designs

Boelens [44] has pointed out that although it is now possible to generate and test arrays of catalysts

> A host of properties and process conditions must be fine-tuned in order to yield a catalyst that is active, selective, and stable. The new high-throughput experimentation (HTE) techniques are invariably based on a two-step approach comprised of primary screening of a large number of candidates (discovery stage) followed by optimization of a small number of leads. In the discovery stage, thousands or even tens of thousands of catalysts may be tested, but often only one binary parameter is scanned (for example, the product yield may be measured only once for each reaction vessel). Thus, good catalysts may be overlooked and not pass into the optimization stage if they score low, for any reason, in the initial discovery tests, and vice versa.

His response is to find actual kinetic parameters for an array of catalysts. Since the analyzer is usually the bottleneck in the process, he has generated a sampling strategy to maximize the amount of information gained with each sample from the array. This strategy selects a small number of optimal sampling points for each reaction [45]. In a subsequent paper the methodology was extended from simple first-order reactions to cascade reactions [46]. It was then generalized using the "T-optimality and D-optimality criteria using a Pareto approach to obtain a sampling design good for both model selection and reaction rate estimation" [47].

2.7.2. Data Cubes

Hendershot has pointed out "the true power of the combinatorial approach, however, can only be realized with the ability to perform quantitative studies

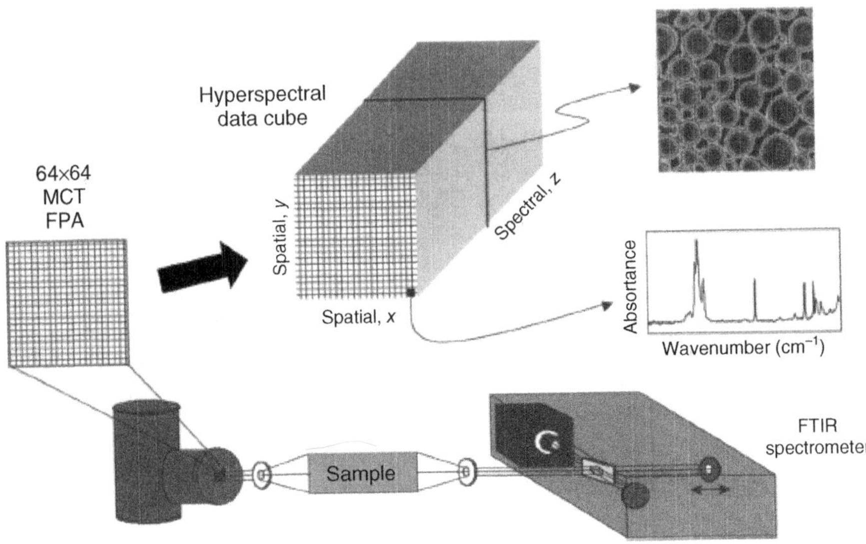

Figure 2.12. FTIR imaging system showing data cube. [Reprinted, with permission, from Snively, C. M., Oskarsdottir, G., and Lauterbach, J., *Catal. Today* **67** (2001): copyright 2003 © Elsevier.]

in parallel. This methodology, in combination with microkinetic modeling of the quantitative data, can take our understanding of heterogeneously catalyzed reactions to a higher level, accelerate discovery of novel catalyst formulations, and ultimately lead to rational catalyst design" [48]. He developed a remarkable method of obtaining detailed kinetic information with high-throughput methodology by combining parallel reactors (currently 16, but expandable) with truly parallel, high time resolution analysis using FTIR (Fourier transform infrared methods). The result is a *data cube* (Fig. 2.12) of gigabyte proportions, which can be quickly processed to a high-resolution view of transient catalyst behavior for multiple catalysts. This has been combined with a knowledge extraction algorithm for determining the most probable elementary steps and rate constants for relatively complex processes [23].

2.8. COMBINATORICS

When the number of possible combinations qualitative factors and their levels becomes astronomical, more brute-force methods can still produce a useful outcome. Hulliger has reported the "single-sample concept," in which a random mixture of a very large number of tiny grains of starting materials is sintered together. He argues that since there are 10^9–10^{12} grains in a 1-cm^3 sample, it is statistically certain that all possible phase diagrams of up to 6 elements will be generated even if there are as many as 40 distinct elements

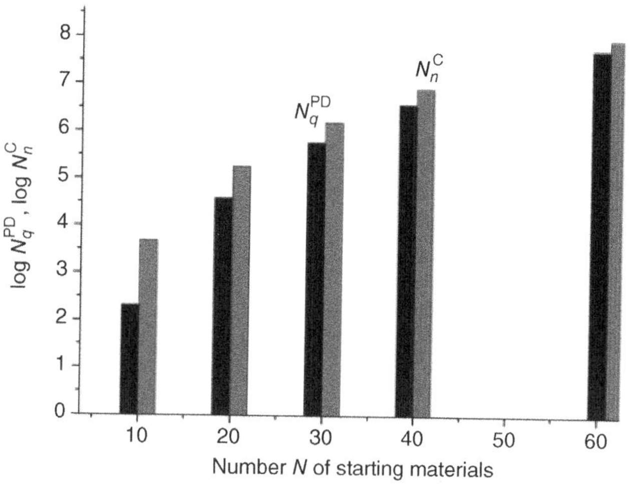

Figure 2.13. Calculated number of possible phase diagrams and local configurations for N starting components and $n = 6$ elements in product grains. [Reprinted, with permission, from Hulliger, J. and Awan, M. A., *Chemistry* (a European journal) **10**(19) (2004); copyright 2004 © Wiley-VCH.]

in the mixture (Fig. 2.13). The key requirement is an exquisitely sensitive analytical method for locating the leads produced. In his work, sintered libraries of 17, 25, and 30 metal oxides were produced. Each was ball-milled to release the combinations produced and passed through a magnetic separation column where the small number of grains showing strong magnetic effects were separated [49,50].

2.9. HOLOGRAPHIC STRATEGY

Vegvari and Tompos advocate a novel method called a "holographic research strategy" (HRS) [51], which is based on an original and ingenious method of visualization of multidimensional experimental data. The experimental space studied is the same continuous multidimensional space examined in Section 2.6. The assumption is that each of N experimental variables can potentially be examined at a finite number m_N of parameter levels. If $m = 2$, we have the common factorial design; for $m = 3$, we can use standard response surface designs like central composites. It is when $m > 3$ that both the design and visualization becomes difficult. Even a moderate case where $N = 6$ and $m = 4$ results in $4^6 = 4096$ points.

Visualization of this type of data uses an inspired arrangement in which one-half of the predictor variables are placed along the X axis (abscissa) of a graph and the other half along the Y axis (ordinate). The points are arranged so that each forms a wave of different periodicity (Fig. 2.14). The results are

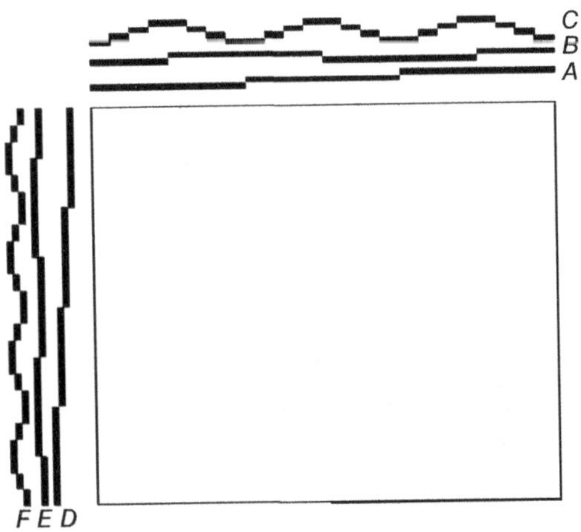

Figure 2.14. Two-dimensional presentation of a continuous six-dimensional experimental space.

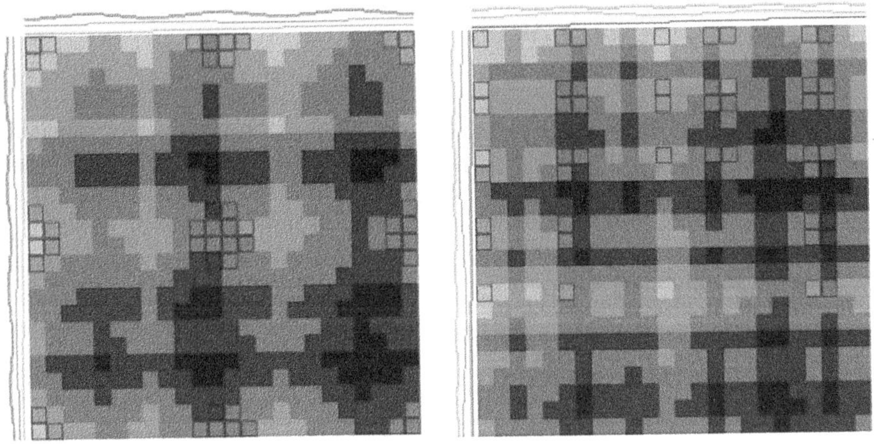

Figure 2.15. Visualization of of a $2^2 \times 3^2 \times 4^2 = 576$-point space with color encoding and one "variable position change." [Reprinted, with permission, from Vegvari, L. et al., *Catal. Today* **81**(3) (2003); copyright 2003 © Elsevier B. V.]

placed in this X–Y space coded by color or gray-scale intensity. This arrangement forms a data-rich graphic, which can easily allow visualization of several hundred thousand points! For example, the eye can easily view and resolve a 200×200 mm grid at 0.5 mm pitch, for a total of 160,000 points. The graphic has the further advantage that the information on the graph can be clustered and declustered by changing the position and periodicity of the predictor variables (Fig. 2.15). This allows quick and flexibly visualization of the association of results with predictor variables. The entire system has been incorporated in a software package [52].

Experimental planning using HRS uses the clustering/declustering method to search for optima. An initial set of points is generated, either with chemical knowledge, at random or in a symmetric pattern. After the experiments are completed, the best point(s) is (are) selected and the array is declustered. The next set of experiments is performed in the (new) immediate neighborhood of the best-points. Successive iterations of experimentation, best-point selection, and declustering effectively sample the neighborhood of the best points in all dimensions. In a demonstration using a virtual space of ~60,000 experimental points, the HRS found the optimum in 9 iterations constituting 153 runs.

The authors compare the HRS with genetic algorithms as effective strategies for exploring continuous multidimensional experimental space. HRS is an entirely deterministic method; once the system parameters are selected, the route of optimization is unequivocally determined. It has the advantage of having the holographic array as a powerful visualization tool of the space. The geometric arrangement in the hologram partially reveals the composition–activity relationship. It has a potential disadvantage of becoming "caught" on a local optimum and missing a global optimum. The stochastic character of a GA will probably cause slower optimization but (particularly with mutations) gives a better chance of escaping a local optimum in favor of a global optimum.

2.10. SPLIT AND POOL METHODS

After its early exuberant development, combinatorial chemistry in the pharmaceutical industry has largely converged to the use of parallel, split and pool methods in synthesis of lead molecules [53]. This methodology uses arrays of synthetic reactions acting on a scaffold molecule typically attached to a polymer bead, with mixing and splitting at intermediate steps, to generate exponentially increasing numbers of discrete products (Fig. 2.16). Since combinatorial materials development does not typically generate single molecular species as its targets, the use of split and pool methods has seen little use.

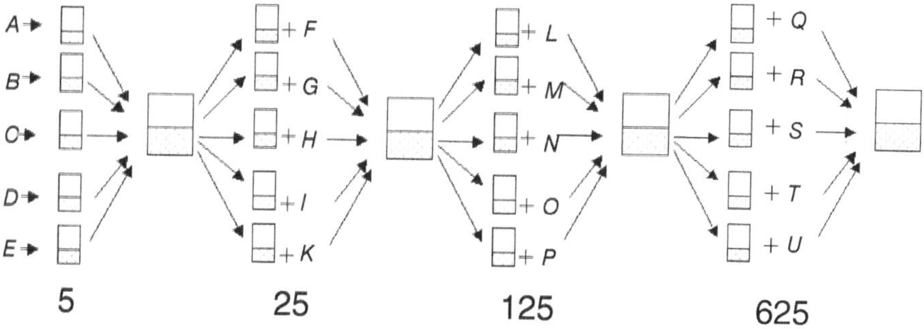

Figure 2.16. Split-and-pool methodology.

Figure 2.17. Single-bead reactor. (Photo courtesy of h.t.e Corporation.)

More recently, however, Mallouk has shown that impregnation of potentially catalytic inorganic materials into alumina beads is sufficiently irreversible that a sequential split and pool impregnation process can generate a useful library [54]. He demonstrated that the beads can be labeled with fluorescent dyes as tags, and that typical catalyst precursors can be loaded on the beads with relatively little leaching and cross-contamination. He points out

> One problem with the split-pool technique is that a large number of beads must be used to ensure that the whole range of compositions is represented by at least one bead. A random library generates many redundant beads. A more economical library, with fewer redundancies, can be designed by directed sorting. This method involves identification of each bead by fluorescence intensity measurements after the second and subsequent split-pool steps.

A similar technique has been described by high-throughput experimentation (HTE), in which the beads are impregnated and then calcined to prevent leaching of catalyst components in subsequent steps. The beads are then loaded into a 16 × 24 parallel single-bead reactor (Fig. 2.17) and analyzed by XRF for elemental composition before testing for catalyst activity. They found that a 3× redundancy of the library was sufficient to ensure complete representation of all possible combinations [55].

2.11. FORMULATION/PROCESS DESIGNS

As the use of high-throughput methods in catalyst research has matured, it has been accompanied by the realization that catalysts are highly sensitive to process conditions, so a simple formulation experiment is highly likely to miss valid catalyst leads. Tibiletti [56] notes:

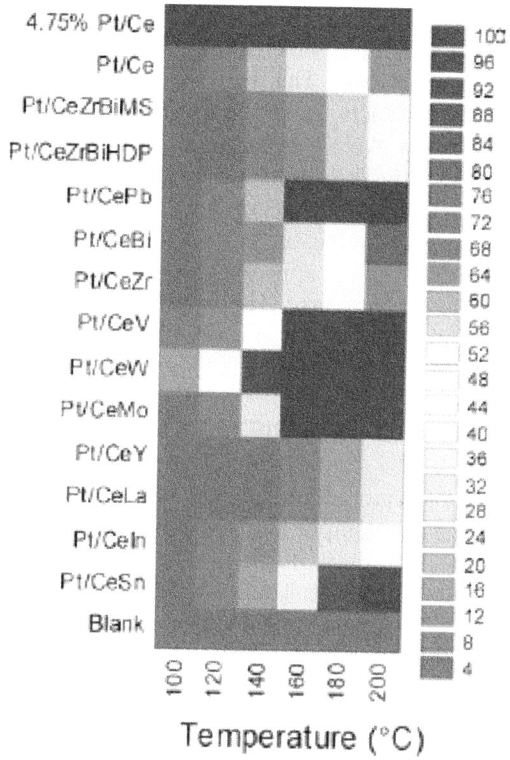

Figure 2.18. Carbon monoxide oxidation in the absence of hydrogen: CO conversion (%) as a function of temperature (°C). [Reprinted, with permission, from Tibiletti, D. et al., *J. Catal.* **225**(2) (2004); copyright 2004 © Elsevier, Inc.]

> In the traditional combinatorial approach it is generally considered that selection of performing elemental combinations can be performed in the primary screening where testing conditions are often simplified to speed up the screening process ... This, however can lead to major errors ... the risks for data misinterpretation when high-throughput experiments are carried out without satisfactory control of temperature and pressure conditions are high.

By conducting high-throughput searches of catalysts over a temperature range he was able to select superior catalysts (Fig. 2.18). He also notes that "the bottleneck lies now with the data analysis and processing" (!).

2.12. ARTIFICIAL NEURAL NETWORKS

The use of artificial neural networks (ANNs) in a high-throughput experimental strategy continues to be an area of great interest [2,57–67]. Baumes points out that in a combinatorial material program a "large number of experiments has to be carried out to highlight high orders of interactions ... most of the experiments (up to 99%) are carried out in irrelevant areas of the search space" [65]. This leads to better strategies that favor a learning process in which each set of experiments is guided by the results of previous ones. He favors ANNs integrated with an evolutionary strategy, because "ANN used at the browsing stage (preliminary screening) poorly predicts the quantitative catalyst performance" [65]. Instead, the ANN is used as a classifier tool in the course of an evolutionary strategy. Furthermore, ANN appears to be much more appropriate as a classifier than as a tool for predicting continuous performance. The latter is probably is probably best done by more conventional regression tools.

2.13. COMMERCIAL AND PROPRIETARY DESIGN METHODS

Several companies have emerged during the past few years to commercialize various aspects of high throughput and combinatorial chemistry. Symyx is the first and best known, but others include Avantium, hte_AG, Novodynamics, Torial, Accelerys, and Inforsense. Their portfolios include various combinations of contract research, consulting, hardware, and informatics.

Each informatics package (Table 2.5) addresses the total workflow issue of a high-throughput laboratory (Fig. 2.19). This includes raw-materials registration, experimental design, sample synthesis/formulation, sample preparation and screening, properties analysis, and data analysis. They are built around a powerful database engine, since enormous amounts of data must be stored and managed, and the corresponding data structures are highly complex. A user-friendly input module, interfaces with robots and analytical systems, statistics, and data visualization are typical.

Several of these design modules combine experimental design concepts with various kinds of modeling tools to generate both real and virtual experiments. Avantium calls this "rational screening" [73]. These are frequently descriptor-based designs, where descriptors for the chemical species factor are

TABLE 2.5. Commercial Informatics Packages for High Throughput Materials Research

Vendor	Informatics Package	Materials Focus	Ref.
Symyx	Library Studio®	Catalysts, polymers	[68]
hte	hteSetup™	Catalysts	[69]
Avantium	Data Analysis Package®	Organics, catalysts	[70]
Novodynamics, Inc.	ArborChem™	Zeolite catalysts	[71]
Accelrys	Materials Studio®	Organics, pharmaceuticals	[72]

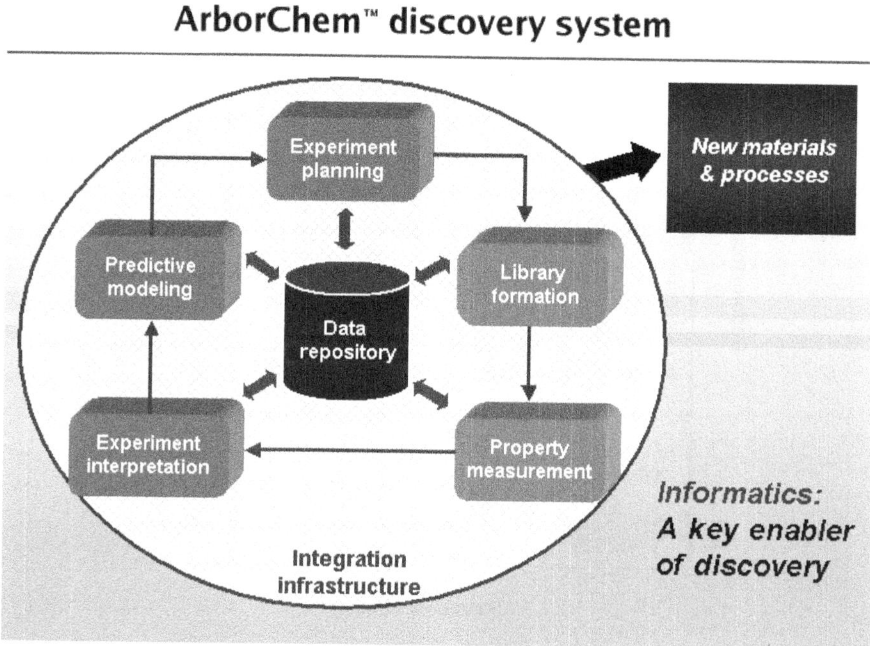

Figure 2.19. An integrated combinatorial chemistry workflow. (Reprinted, with permission, from Novodynamics, Inc.)

generated. These can be constitutional (e.g., molecular weight); topological, geometric, or quantum-chemical. Since many descriptors can be generated for each compound and they are frequently highly correlated, principal-components analysis (PCA) is used to generate a relatively small set of orthogonal properties that can be used as factors in a DOE [5].

Accelrys claims particular effectiveness in high-throughput formulation development through the use of

- *Categorical model analysis*. Accelrys has patented a "partially unified multiple property recursive partitioning" (PUMP-RP) tool for decision-tree predicting of multiple-response categorical data [74].
- *Backpropagation neural network models*, which also predict multiple responses, and can be used with missing response and/or descriptor data.
- *Dual-response system* approaches to robust design, in which the mean is optimized while the standard deviation is minimized [75].
- Genetic algorithms applied to mixture models.

The large amount of data available from HTS is a requirement for effective use of the first two of these tools.

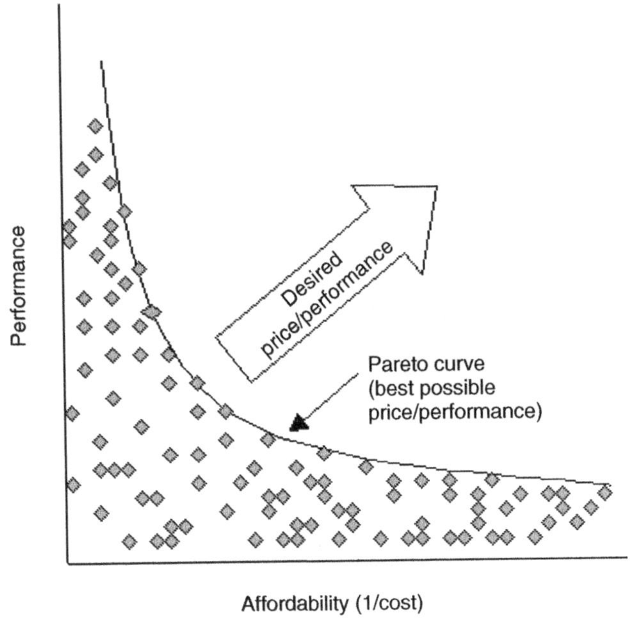

Figure 2.20. Pareto surface.

Considerable effort has also been reported in the use of Pareto optimization. Pareto optimization is the search for a surface in formulation/property space (Fig. 2.20), which bounds a system in which improvement of one (or more) objective results in deterioration of the other objectives [76,77]. These objectives can include both the mean *and* the variance of critical performance properties. The goal is balancing the "multiple performance properties that must be addressed to ensure the success of a commercial material."

Symyx has the most evolved set of software; its Renaissance® Library Studio® is used at many of the major US companies using HTS, including GE, Dow, and Exxon-Mobil. The software is notable in the number of robotic systems that can be directly interfaced with it. Its primary design methodology is essentially one of mixture gradients. The user can easily specify gradients of each component in the mixture across an array of virtually any dimension. The results are intuitively visualized using an array of pie charts (Fig. 2.21).

2.14. DATA PIPELINES

Although not strictly an experimental design tool, data pipelines are likely to have a significant effect on high throughput technologies. These are software systems recently developed and released by several companies such as SciTegic (a subsidiary of Accelerys), Inforsense, and Avantium (Table 2.6). Data

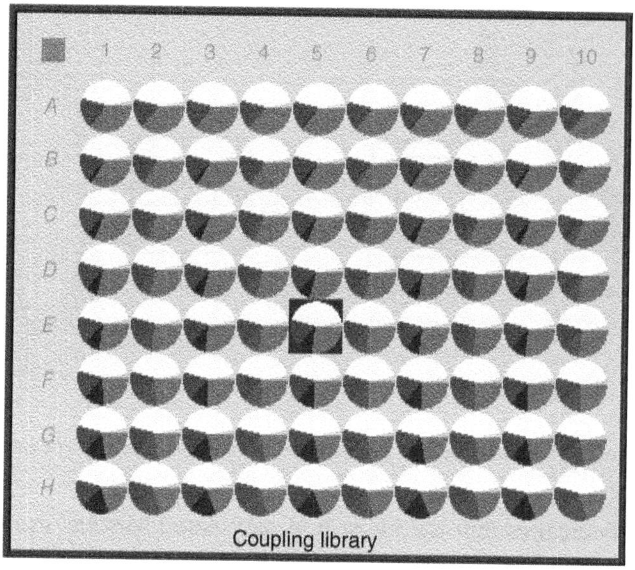

Figure 2.21. Design array composition set up on Renaissance Suite software. (Photo courtesy of Symyx Corporation.)

TABLE 2.6. Data Pipeline Software

Company	Product	Reference
SciTegic	Pipeline Pilot™	http://www.scitegic.com
InforSense	Knowledge Discovery Environment	http://www.inforsense.com/products/
Avantium	Data Analysis Package®	http://www.avantium.com

pipelining is the processing, analysis, and mining of large volumes of data through a user-defined computational protocol. It guides the flow of data through a network of modular computational components, which are typically assembled through a drag-and-drop Windows interface (Fig. 2.22). At the moment the benefits are going to accrue mainly at the downstream end (gathering data from machines, basic data reduction, and data analysis). However, it is being realized that the more upstream activities (experiment design, driving synthesis robots, etc.) will also benefit by data pipelining approaches. Inforsense KDE, for example, has a broad connectivity with MatLab, which makes it possible to access the "design of experiment" capabilities in MatLab's Statistics Toolbox. The specific advantage of the product for design of experiments is that not only does it brings a number of tools together in one application (statistics, text mining, cheminformatics) but once a specific design workflow has been developed, it can be used over and over again using different datasets. Furthermore, the graphical representation of the workflow

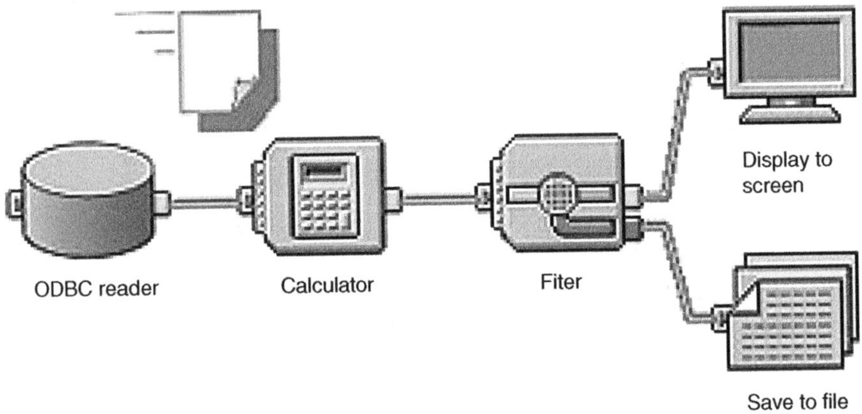

Figure 2.22. Data pipeline. (Photo courtesy of SciTegic Corporation.)

automatically ensures that the know-how associated with the design exercise is documented and not just its outcome (i.e., a set of synthesis candidates or a list of processing instructions) [78].

2.15. THE NEXT CHALLENGES

Design of materials discovery libraries—and, indeed, rapid progress in materials development in general—awaits a methodology to predict (or at least structure) the performance diversity of a library. As of now, *the underpinning for descriptor based preselection of compounds, which are used in the drug industry, the similar property principle, is basically missing when we are dealing with solids* [79]. These authors suggest an ambitious program in which the data from a first screening of intuitively diverse materials are used to find clusters of similar behavior. In parallel, a database of potential descriptors of those materials is generated. These descriptors must include physical properties, electronic structures, and synthetic methods. Correlation of the descriptors and the response clusters would be used to find descriptor vectors that are discriminative with respect to the cluster assignment. Finally, these vectors would be validated and used to generate new libraries, either real or virtual, which will have a higher probability of locating the desired properties in experimental space. The authors do note, "One should not underestimate the problems with respect to the algorithms and data treatment associated with this work"(!)

REFERENCES

1. Cawse, J. N., Experimental strategies for combinatorial and high throughput materials development, *Acc. Chem. Res.* **34**:213–221 (2001).

2. Harmon, L., Experiment planning for combinatorial materials discovery, *J. Mater. Sci.* **38**(22):4479–4485 (2003).
3. Cawse, J. N., ed., *Experimental design for combinatorial and high throughput materials development*, Wiley, New York, 2002, p. 317.
4. Farrusseng, D., Baumes, L., and Mirodatos, C., Data management for combinatorial heterogeneous catalysis: Methodology and development of advanced tools, in *High-Throughput Analysis*, R. A. Potyrailo and E. J. Amis (eds.) Kluwer/Academic Press, 2003, New York, pp. 551–579.
5. McKay, B. et al., Advances in multivariate analysis in pharmaceutical process development, *Curr. Opin. Drug Discov. Devel.* **6**(6):966–977 (2004).
6. Rosso, V. W., Pazdan, J. L., and Venit, J. J., Rapid optimization of the hydrolysis of *N'*-trifluoroacetyl-*S*-*tert*-leucine-*N*-methylamide using high-throughput chemical development techniques, *Org. Process Res. Devel.* **5**(3):294–298 (2001).
7. Cornell, J. A., *Experiments With Mixtures: Designs, Models, and the Analysis of Mixture Data*, 3rd ed., Wiley, New York, 2002, p. 649.
8. Potyrailo, R. A. and Amis, E. J., *High Throughput Analysis: A Tool for Combinatorial Materials Science*, Kluwer Academic/Plenum, New York, 2002.
9. Kennedy, K. et al., Rapid method for determining ternary-alloy phase diagrams, *J. Appl. Phys.* **36**(12):3808–3810 (1965).
10. Hanak, J. J., The multi-sample concept in materials research: Synthesis, compositional analysis and testing of entire multicomponent systems, *J. Mater. Sci.* **5**:964–971 (1970).
11. Schneemeyer, L. F. and Van Dover, R. B., High throughput synthetic approaches for the investigation of inorganic phase space, in *Experimental Design for Combinatorial and High Throughput Materials Development*, J. N. Cawse (ed.), Wiley, New York, 2002.
12. Potyrailo, R. A., Karim, A., Wang, Q., eds., Combinatorial and Artificial Intelligence Methods in Material Science II, MRS Proceedings Vol. 804, Warrendale, PA, 2004.
13. Takahashi, R. et al., Design of combinatorial shadow masks for complete ternary-phase diagramming of solid state materials, *J. Chem. Inform. Comput. Sci.* **6**(1):50–53 (2004).
14. Danielson, E. et al., A combinatorial approach to the discovery and optimization of luminescent materials, *Nature* (Lond.) **389**:944–948 (1997).
15. Potyrailo, R. A. et al., High-throughput fabrication, performance testing, and characterization of one-dimensional libraries of polymeric compositions, *Macromol. Rapid Commun.* **24**(1):123–130 (2003).
16. Reddington, E. et al., Combinatorial electrochemistry: A highly parallel, optical screening method for discovery of better electrocatalysts, *Science* **280**:1735–1737 (1998).
17. Sun, Y., Buck, H., and Mallouk, T. E., Combinatorial discovery of alloy electrocatalysts for amperometric glucose sensors, *Anal. Chem.* **73**(7):1599–1604 (2001).
18. Mallouk, T. E. and Smotkin, E. S., Combinatorial catalyst development methods, in *Handbook of Fuel Cells—Fundamentals, Technology and Applications*, W. Vielstich, A. Lamm, and H. Gasteiger (eds.), Wiley, Chichester, UK, 2002, pp. 334–347.

19. Cawse, J. N. and Wroczynski, R., Combinatorial materials development using gradient arrays: Designs for efficient use of experimental resources, in *Experimental Design for Combinatorial and High Throughput Materials Development*, J. N. Cawse (ed.), Wiley, New York, 2002, pp. 109–128.

20. Cassell, A. M. et al., High throughput methodology for carbon nanomaterials discovery and optimization, *Appl. Catal. A: General* **254**:85–96 (2003).

21. Wolf, D. and Baerns, M., An evolutionary strategy for the design and evaluation of high throughput experiments, in *Experimental Design for Combinatorial and High Throughput Materials Development*, J. N. Cawse (ed.), Wiley, New York, 2002, pp. 147–162.

22. Wright, T. et al., Optimizing the size and configuration of combinatorial libraries, *J. Chem. Inform. Comput. Sci.* **43**(2):381–390 (2002).

23. Caruthers, J. M. et al., Catalyst design: Knowledge extractions from high-throughput experimentation, *J. Catal.* **216**:98–109 (2003).

24. Serra, J. M. et al., Styrene from toluene by combinatorial catalysis, *Catal. Today* **81**(3):425–436 (2003).

25. Sohn, K.-S., Lee, J. M., and Shin, N., A search for new red phosphors using a computational evolutionary optimization process, *Adv. Mater.* **15**(24):2081–2084 (2003).

26. Kirsten, G. and Maier, W. F., Strategies for the discovery of new catalysts with combinatorial chemistry, *Appl. Surf. Sci.* **223**(1–3):87–101 (2004).

27. Cawse, J. N., Baerns, M., and Holena, M., Efficient discovery of nonlinear dependencies in a combinatorial catalyst data set, *J. Chem. Inform. Comput. Sci.* **44**:143–146 (2004).

28. Opticat Ver 0.92, Frédéric Clerc, Laval, France, 2007.

29. Pereira, S. R. M. et al., Effect of the genetic algorithm parameters on the optimization of heterogeneous catalysts, *QSAR Combin. Sci.* **24**(1):45–57 (2005).

30. Reardon, B. J. and Bingert, S. R., Inversion of tantalum micromechanical powder consolidation and sintering models using Bayesian inference and genetic algorithms. *Acta Materialia* **28**:647–658 (2000).

31. Reetz, M. T. and Jaeger, K. E., Enantioselective enzymes for organic synthesis created by directed evolution, *Chemistry* (a European journal) **6**:407–412 (2000).

32. Reetz, M. T., Combinatorial and evolution-based methods in the creation of enantioselective catalysts, *Angew. Chem. Int. Ed.* **40**:284–310 (2001).

33. Kunsch, H. R., Agrell, E., and Hamprecht, F. A., Optimal lattices for sampling, *IEEE Trans. Inform. Theory* **51**(2):23 (2005).

34. Hamprecht, F. A. and Agrell, E., Exploring a space of materials: Spatial sampling design and subset selection, in *Experimental Design for Combinatorial and High Throughput Materials Development*, J. N. Cawse (ed.), Wiley, New York, pp. 277–308 (2002).

35. Conway, J. H. and Sloane, N. J. A., *Sphere packings, lattices, and groups*, 2nd ed., Grundlehren der mathematischen Wissenchaften, Vol. 290, Springer-Verlag, New York, 1988.

36. Dixon, J. M. et al., An experiment planner for performing successive focused grid searches with an automated chemistry workstation, *Chemometr. Intell. Lab. Syst.* **62**:115–128 (2002).

37. Matsumoto, T., Du, H., and Lindsey, J. S., A parallel simplex search method for use with and automated chmeistry workstation, *Chemometr. Intell. Lab. Syst.* **62**:129–147 (2002).

38. Matsumoto, T., Du, H., and Lindsey, J. S., A two-tiered strategy for simplex and multidirectional optimization of reactions with an automated chemistry workstation, *Chemometr. Intell. Lab. Syst.* **62**(2):149–158 (2002).

39. Du, H. and Lindsey, J. S., An approach for parallel and adaptive screening of discrete compounds followed by reaction optimization using an automated chemistry workstation, *Chemometr. Intell. Lab. Syst.* **62**(2):159–170 (2002).

40. Gosset, AT&T Shannon Labs, Florham Park, N.J., 2003. http://www.research.att.com/~njas/gosset/index.html#CO

41. Royle, J. A. and Nychka, D., An algorithm for the construction of spatial coverage designs with implementation in S-Plus, *Comput. Geosci.* **24**(5):479–488 (1998).

42. Fang, K.-T. and Lin, D. K. J., Uniform Designs website, 2003.

43. Fang, K.-T. and Lin, D. K. J., *Uniform Experimental Designs and Their Applications in Industry*, Hong Kong Baptist University, Hong Kong, 2001, pp. 1–40.

44. Boelens, H. F. M. et al., Tracking chemical kinetics in high-throughput systems, *Chemistry* (a European journal) **9**(16):3876–3881 (2003).

45. Rothenberg, G., Optimal on-line sampling of parallel reactions: general concept and a specific spectroscopic example, *Catalysis Today* **81**:359–367 (2003).

46. Iron, D. et al., Kinetic studies of cascade reactions in high-throughput systems, *Anal. Chem.* **75**(23):6701–6707 (2003).

47. Westerhuis, J. A. et al., Model selection and optimal sampling in high-throughput experimentation, *Anal. Chem.* **76**(11):3171–3178 (2004).

48. Hendershot, R. J. et al., High-throughput catalytic science: Parallel analysis of transients in catalytic reactions. *Angew. Chem. Int. Ed.* **42**(10):1152–1155 (2003).

49. Hulliger, J. and Awan, M. A., "Single sample concept": Theoretical model for a combinatorial approach to solid-state inorganic materials, *J. Combin. Chem.* **7**(1): 73–77 (2004).

50. Hulliger, J. and Awan, M. A., A single-sample concept (SSC): A new approach to the combinatorial chemistry of inorganic materials, *Chemistry* (a European journal) **10**(19):4694–4702 (2004).

51. Vegvari, L. et al., Holographic research strategy for catalyst library design: Description of a new powerful optimisation method, *Catal. Today* **81**(3):517–527 (2003).

52. Xhrs for Linux, Meditor General Information Bureau, Zrinyi, Hungary, 2005.

53. Coffen, D. L. and Luithle, J. E. A., Introduction to combinatorial chemistry, in *Handbook of Combinatorial Chemistry*, K. C. Nicolaou, R. Hanko, and W. Hartwig (eds.), Wiley-VCH, Weinheim, 2002, pp. 10–23.

54. Sun, Y. et al., The split-pool method for synthesis of solid-state material libraries, *J. Combin. Chem.* **4**:569–575 (2002).

55. Klein, J. et al., Application of a novel split&pool-principle for the fully combinatorial synthesis of functional inorganic materials, *Appl. Catal. A: General* **254**:121–131 (2003).

56. Tibiletti, D. et al., Selective CO oxidation in the presence of hydrogen: Fast parallel screening and mechanistic studies on ceria-based catalysts, *J. Catal.* **225**(2):489–497 (2004).
57. Corma, A. et al., Application of artificial neural networks to combinatorial catalysis: Modeling and predicting ODHE catalysts, *Chemphyschem* **3**:939–945 (2002).
58. Jurs, P. C. and Mattioni, B. E., Prediction of glass transition temperatures from monomer and repeat unit structure using computational neural networks, *J. Chem. Inform. Comput. Sci.* **42**:232–240 (2002).
59. Mosier, P. D. and Jurs, P. C., QSAR/QSRP studies using probabilistic neural networks and generalized regression neural networks, *J. Chem. Inform. Comput. Sci.* **42**:1460–1470 (2002).
60. Holena, M. and Baerns, M., Feedforward neural networks in catalysis: A tool for the approximation of the dependency of yield on catalyst composition, and for knowledge extraction, *Catal. Today* **81**:485–494 (2003).
61. Serra, J. M. et al., Can artificial neural networks help the experimentation in catalysis? *Catal. Today* **81**(3):393–403 (2003).
62. Serra, J. M. et al., Neural networks for modelling of kinetic reaction data applicable to catalyst scale up and process control and optimisation in the frame of combinatorial catalysis, *Appl. Catal. A: General* **254**(1):133–145 (2003).
63. Tompos, A. et al., Information mining using artificial neural networks and "holographic research strategy," *Appl. Catal. A: General* **254**(1):161–168 (2003).
64. Adams, N. and Schubert, U. S., From data to knowledge: chemical data management, data mining, and modeling in polymer science, *J. Combin. Chem.* **6**(1):12–23 (2004).
65. Baumes, L. et al., Using artificial neural networks to boost high-throughput discovery in heterogeneous catalysis, *QSAR Combin. Sci.* **23**(9):767–778 (2004).
66. Burello, E., Farrusseng, D., and Rothenberg, G., Combinatorial explosion in homogeneous catalysis: Screening 60,000 cross-coupling reactions, *Adv. Synth. Catal.* **346**(13–15):1844–1853 (2004).
67. Klanner, C. et al., The development of descriptors for solids: Teaching "Catalytic intuition" to a computer, *Angew. Chem. Int. Ed.* **43**(40):5347–5349 (2004).
68. Library Studio®, Symyx Corporation, Santa Clara, CA, 2005.
69. hteSetup™ and hteControl™, hte Aktiengesellschaft, Heidelberg, Germany, 2005.
70. Data Analysis Package®, Avantium Technologies, Amsterdam, the Netherlands, 2005.
71. NovoD Xplanner®, NovoDynamics, Ann Arbor, MI, 2005.
72. Materials Studio 4.1®, Accelrys Corporation, San Diego, CA, 2005.
73. Verspui, G. et al., Rational screening: Parallel experimentation and predictive modeling applied to chemical process R&D, *PharmaChem* **4**:2–6 (2004).
74. Stockfisch, T. P., *Methods and Systems of Classifying Multiple Properties Simultaneously Using a Decision Tree*, Accelrys, UK, 2002, p. 45.

75. Koksoy, O. and Doganaksoy, N., Joint optimization of mean and standard deviation using response surface models, *J. Qual. Technol.* **35**(3):239–252 (2003).

76. Nicolaides, D., Robust material design: A new workflow for high-throughput experimentation and analysis, *QSAR Combin. Sci.* **24**(1):7–14 (2005).

77. Doyle, M. J., Research and development challenges, in *Combi 2004*, Knowledge Foundation, Arlington, VA, 2004.

78. Adams, N. and Schubert, U. S., Personal communication, 2005.

79. Klanner, C. et al., How to design diverse libraries of solid catalysts? *QSAR Combin. Sci.* **22**(7):729–736 (2003).

CHAPTER 3

Polymeric Discrete Libraries for High-Throughput Materials Science: Conventional and Microfluidic Library Fabrication and Synthesis

KATHRYN L. BEERS and BRANDON M. VOGEL

Polymers Division
National Institute of Standards and Technology (NIST)
Gaithersburg, Maryland[1,2]

3.1. INTRODUCTION

High-throughput and combinatorial synthesis find their beginnings in the pharmaceutical industry because structure–activity relationships are difficult to predict and more easily identified by empirical methods. These methods find application in biological assays and biochemical synthesis, since the parameter space involved with biological processes is enormous due to the complex network of molecules involved in directing the "synthesis" of peptides, proteins, and other factors. In addition, inorganic chemistry provides model systems to conduct parallel synthesis and screening because compositional libraries are simply prepared by sequential thin-film deposition of individual or mixed components.

However, polymeric discrete libraries are difficult synthesize, because many reactions do not go to completion, resulting in impurities that may affect the

[1] Certain equipment, instruments, or materials are identified in this chapter to adequately specify the experimental details. Such identification does not imply recommendation by the National Institute of Standards and Technology, nor does it imply the materials are necessarily the best available for the purpose.

[2] Contribution of the National Institute of Standards and Technology, not subject to copyright in the United States.

Combinatorial Materials Science, Edited by Balaji Narasimhan, Surya K. Mallapragada, and Marc D. Porter

physical properties of the materials synthesized. Additionally, there are only a few methods to prepare polymers, and many different types of monomers cannot be combined in a polymerization. For example, a polyester is synthesized from a diacid and a diol in a condensation reaction, but a diacid cannot be condensed with a diacrylate to produce a polymer. Many of these polymerization reactions require catalysts, initiators, controlled atmosphere, or high temperatures. Also, screening polymer libraries in situ is difficult because most testing methods require a polymeric thin film or the polymer dissolved in a solvent. The final problem with discrete library synthesis is the cost of setting up a facility to synthesize, characterize, and handle the informatics involved with data processing. These facilities can cost upward of $5 million when liquid handling systems, parallel synthesis reactors and online chromatography are factored. All of these reasons contribute to the slow progress in discrete polymer library synthesis and characterization in academics. Not until recently have polymer libraries become more common and technically feasible.

In this chapter, we will consider the recent literature on parallel discrete polymer synthesis by conventional methods, such as reactions in wells, and microfluidics, an inexpensive, emerging method to prepare polymer libraries. A review of the literature concerning biomaterials, polymers as catalysts, and functional polymers and dendrimers is presented. Section 3.3 contains an introduction to microfluidics, fabricating devices, past research in the field related to polymer synthesis, and future directions. The field of microfluidics is only about a decade old but may revolutionize the methods of parallel synthesis and characterization because of the ease of device fabrication, small material volumes, and low cost. We close this review with a summary and an outlook on parallel synthesis. We point the reader to a couple of detailed reviews on polymer synthesis in the literature for more information on such topics as the equipment involved in preparing conventional libraries [1] and industrial research conducted with parallel synthesis [2].

3.2. CONVENTIONAL DISCRETE POLYMER LIBRARIES

3.2.1. Biomaterial Libraries

Biomaterial libraries are some of the most successful examples of combinatorial chemistry. Their success is a result of the infrastructure in place from the wealth of biological assays designed for high-throughput screening. The goal of any combinatorial study should be to correlate the primary structure of a designed library to the function of the prepared polymers and perhaps use the information to select or design new polymers for an application. Some common properties to correlate structure with are thermal properties (e.g., glass transition (T_g), crystalline melting point (T_m), mechanical properties (modulus), or wetting properties (contact angle). One of the early papers discussing parallel polymer synthesis is from Kohn and coworkers describing a library of 112 polyarylates derived from 14 distinct tyrosine diphenols and eight aliphatic diacids (Fig. 3.1) [3]. These polymers were synthesized on a 0.2 g scale within individual reaction vessels set up in a water shaker bath. The resulting alternating copolymers were biodegradable carbonates with potential applications in drug delivery and tissue engineering. All the polymers were characterized by gel permeation chromatography (GPC), differential scanning calorimetry (DSC), and water contact angle. Each of the polymers synthesized had relative molecular masses above 50,000 g/mol and polydispersities between 1.4 and 2.0. The authors correlated the glass transition temperature to the chemical structure of the polymer. The T_g values of the polymers decreased from 91°C to 2°C in one-degree increments as a function of the presence of oxygen or number of carbons in the polymer backbone and sidechains. The water contact angle followed a similar trend decreasing from 101°C to 64°C with increasing chain length. One of the important contributions of this work was to assign a flexibility index, related to the structure of the polymers synthesized. The *flexibility index* is the total number of carbons in the aliphatic segments of the backbone and the sidechains of a polymer. For polymers without any oxygen atoms, the water contact angle and glass transition temperature strongly correlated to the flexibility index. Cell contact assays were performed on a subset of 42 polymers. They found that cell growth was not a strong function of hydrophobicity, but rather the presence of oxygen in the polymer backbone.

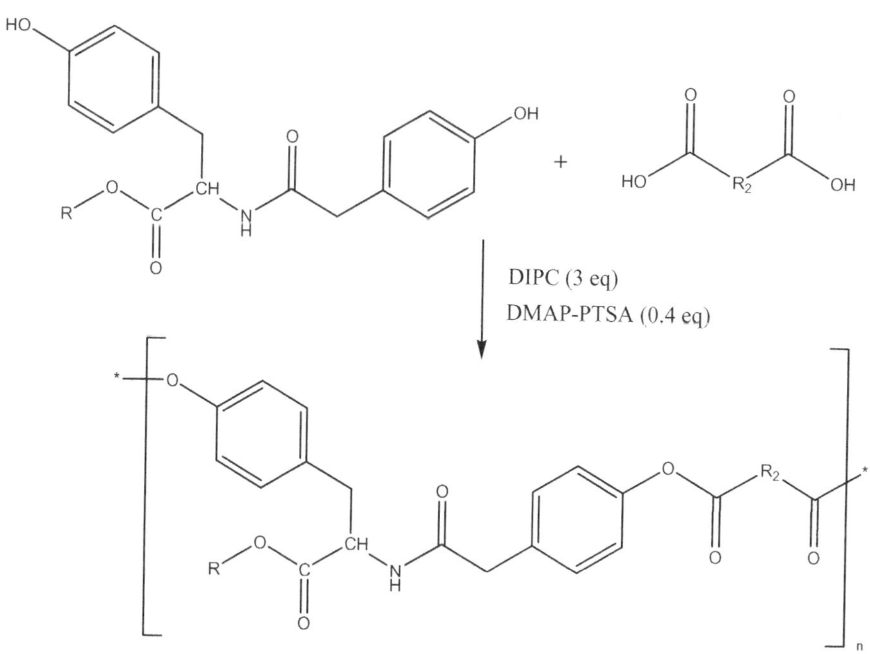

Figure 3.1. Polymerization scheme for the synthesis of tyrosine-derived polycarbonates.

In a follow-up study, a target library of 72 polyarylates were synthesized to gain insight into the effect of substitution on material properties and cell proliferation [4]. Kohn and coworkers found that increasing hydrophobicity decreased cell proliferation except for polymers that contained oxygen, as noted before. However, the authors systematically varied the oxygen content by making a polymer from a monomer with one oxygen (diglycolate) and a polymer from a monomer with no oxygen (diglutarate) in the backbone. Their physical properties were nearly the same, including contact angle, but cells proliferated on the oxygen-containing polymer and not on the all-aliphatic polymer. The authors note that several other studies tried to relate contact angle to cell proliferation but many of the results in the literature were contradictory. This study highlights an important advantage of combinatorial methods—the ability to systematically vary properties to discover relationships that would otherwise not be clear.

Brocchini and coworkers synthesized a series of 16 polyesters derived from serinol [5]. The polymers were screened by water contact angle, thermal properties, and cell attachment assays. The authors tried to create a series of polymers that would yield better physical and mechanical properties than poly(glycolic acid) (PGA) because of PGA's notoriously poor solubility in common solvents. However, some difficulty was met in synthesizing the desired polyesters because of low reaction yields and nonsolid polymers. The polymers were synthesized by reacting a commercially available series of aliphatic diacids with *N*-substituted serinols in the presence of an esterification catalyst. The *N*-substituted serinols provided a means to systematically vary the sidechain functionality while the diacids were used to change the mainchain characteristics. The authors note that little cell attachment occurred on films of those polyesters that contained a hexyl sidechain or advancing water contact angles greater than 74°.

More recently, Langer and coworkers developed a chemistry that is suitable for parallel solution polymerization [6,7]. By simply mixing a diacrylate and a primary amine or secondary diamine, the authors were able to produce a library of poly(β-aminoesters) that are hydrolytically degradable (Fig. 3.2). They designed their library to optimize transfection vectors (polymers that are able to collapse DNA and deliver it to cells). Transfection is an important method to manipulate cellular function that has consequences in many therapies, including cancer. The reactions scale of the reactions (0.6–0.8 g), provided just enough material for characterization. An early screen of water solubility removed nearly half of the candidate polymers as they were insoluble at a usable concentration for the transfection assays. However, the authors found that hydrophobic character was necessary to transfect DNA into cells. Polymers containing histidine units resulted in transfection efficiencies 4–8 times higher than commercially available systems. The authors showed that polymer/DNA transfection complexes less than 250 nm in diameter were necessary for cellular uptake. Also, the ratio of polymer nitrogen to the DNA phosphorous (N/P) was an important parameter in internalization of the

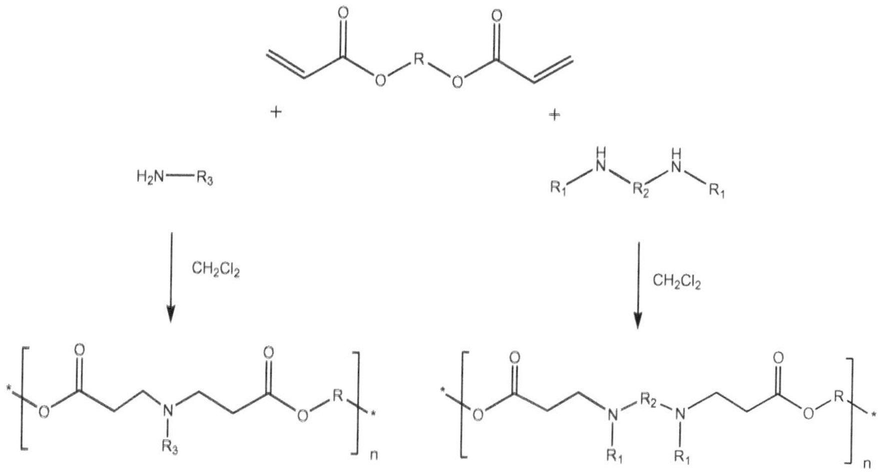

Figure 3.2. Synthetic reaction scheme to produce poly(β-aminoesters).

polymer/DNA complexes. The results suggest an N/P ratio of 20 : 1 was not high enough for internalization but a ratio of 40 : 1 increased the surface charge, resulting in a positive zeta potential (measurement of surface charge) and better internalization and transfection efficiency. The structure of polymers with a 40 : 1 N/P had two amines per repeat unit.

In a later study, Langer and coworkers automated the synthesis of aminoesters and could perform 1000 reactions in a day [8]. Of a 2350-member library, several common features became clear for the design of successful transfection vectors. The presence of monomers with mono or dialcohol sidegroups and bis(secondary amines) were important to high transfection rates. Further publications on the same system of aminoester polymers discussed the importance of relative molecular mass and endgroups to transfection efficiency [9,10]. Langer and coworkers found that a relative molecular mass greater than 10,000 g/mol was important for transfection. Moreover, because of the condensation polymerization mechanism used to synthesize these polymers, diacrylate-amine stoichiometry controlled the endgroup functionality. An excess of diacrylate resulted in acrylate endgroups on the polymer and an amine-terminated polymer resulted from adding excess amine to the reactants. The biological characterization showed that acrylates performed worse than amine-terminated endgroups during transfection. The exact reason for this is unknown, but the authors hypothesize that either a reaction between DNA and an acrylate occurs or there is greater surface charge with the amine-terminated polymers.

This series of papers is a good example of the power of high-throughput experimentation. By creating focused libraries with variation in structure or relative molecular mass or surface charge, Langer and coworkers were able to explore systematically the influence of each of these key parameters on the ability of their cationic polymers to transfect DNA. All of these studies culminated in the in vivo testing of a few candidate cationic polymers with tumor containing mice. While the commercial polyethylene amine transfection vector results in toxicity to healthy tissue, the candidate vector optimized by high-throughput methods had little to no cytotoxicity and resulted in a 40% reduction in the animal's tumor size [10]. Although transfection vectors were an ideal system to facilitate biomaterials synthesis and biological screening, the literature is not limited to this area. Another area of intense research is the use of arrays of biomaterials to screen cell interactions and drug release.

Vogel and coworkers used parallel synthesis and high-throughput characterization to screen a series of polyanhydrides for drug delivery [11]. Polyanhydrides are a Federal Drug Administration (FDA)-approved biomaterial for controlled drug delivery. The polymers are synthesized from acetylated diacid prepolymers by polycondensation at 180°C under low vacuum. By varying the copolymer composition, one can tune polymer degradation from literally hours (short aliphatic diacids) to years (aromatic diacids). As a result, the parameter space for determining the appropriate copolymer composition to not only degrade over the time period of interest but also stabilize the therapeutic is enormous. The authors fabricated microwells using rapid prototyping contact lithography of the optical adhesive NOA-81 on glass slides. Polymer libraries (Fig. 3.3) were generated by depositing prepolymers into the wells and allowing the carrier solvent to dry before bulk polymerization of all 100 samples in parallel in a vacuum oven. Figure 3.4 shows an image of the copolymer library and the library imaged using an infrared (IR) microscope. The IR map displays the gradient in copolymer composition from aliphatic polymer poly(sebacic anhydride) poly(SA) to aromatic containing copolymer poly(1,6-(bis-*p*-carboxyphenoxy)hexane) (poly(CPH)) (blue to red false color). The individual IR spectra were quantified to provide the composition within each well. The authors also sampled some of the wells and performed nuclear magnetic resonance (NMR) to determine monomer sequence lengths and compositions, which also agreed with previously reported data. Rapid prototyping was also used to screen the polyanhydride copolymer release characteristics.

Using microdissolution wells, also fabricated from rapid prototyping, five copolymers replicated 5 times were synthesized with the nonreactive ultravio-

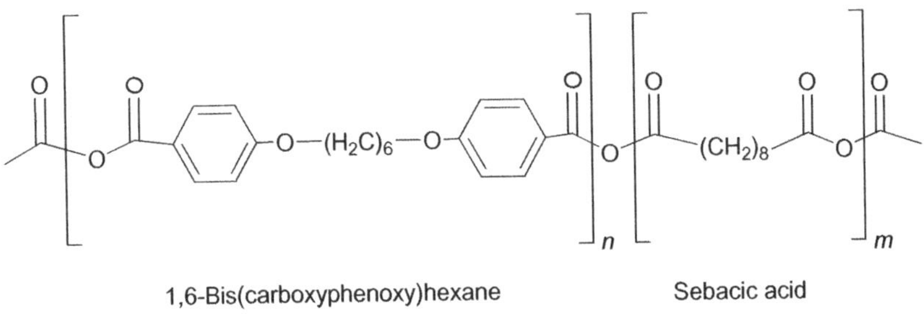

Figure 3.3. Polyanhydride poly 1,6-(bis-*p*-carboxyphenoxy)hexane-co-sebacic anhydride) (polyCPH-co-SA).

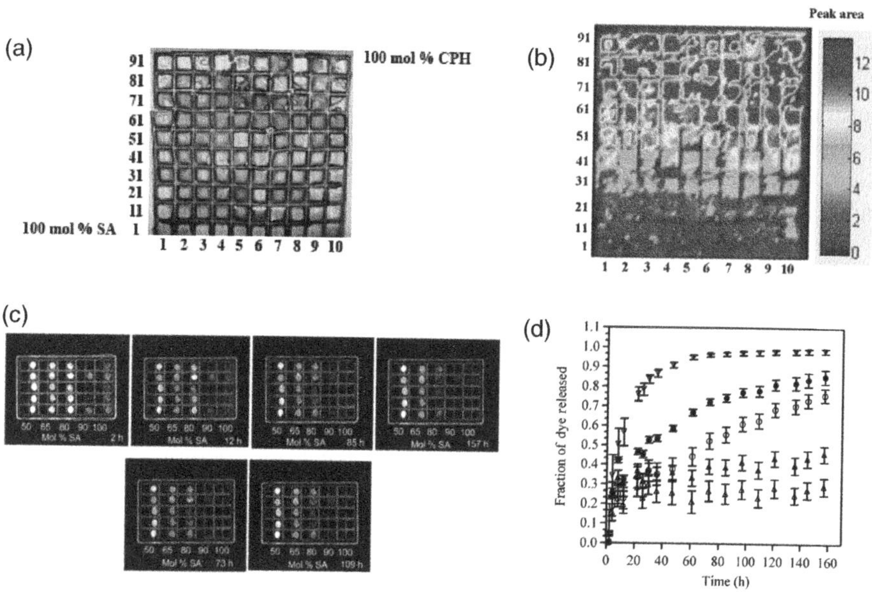

Figure 3.4. (a) Image of a poly(CPH-co-SA) library; (b) IR map of the same library as 4a blue denotes high SA compositions, red high CPH fractions; (c) time series of images taken to screen drug dissolution rate from five copolyanhydrides (0, 10, 20, 50 mol% CPH); (d) Extracted data from the time series showing release kinetics for the five copolymers mentioned previously (top–bottom progression indicates increasing CPH content). (Images taken from Ref. 9.)

let dye ethidium bromide bisacrylamide (EB). The dissolution apparatus was filled with phosphate–buffered saline (PBS) and enclosed with a conformal piece of polydimethylsiloxane and clamps. The release of ethidium bromide bisacrylamide was followed by tracking the decay in the fluorescence intensity in the polymer films with a charge-coupled device (CCD) camera and quantified by image analysis. This technique is a nonsampling method to quantify release kinetics. Figure 3.4c shows several different timepoints and the loss in fluorescence signal with time. Figure 3.4d shows accumulated data for five different copolymers of poly(CPH-co-SA) (from top to bottom the values are increasing in CPH content 0, 10, 20, 35, and 50 mol% CPH). Notice the low CPH content releases a burst of EB and increasing the CPH content slows the burst effect (indicative of stabilization). The 20 : 80 CPH : SA composition gave near-zero-order release kinetics of EB when screened with the microwell apparatus, which is the same composition that is FDA approved for release of hydrophobic chemotherapy drugs. It should be pointed out that a typical dissolution experiment using a United States Pharmacopoeia dissolution apparatus uses six to eight tanks filled with 1 L of PBS and a 0.1 g tablet of polymer containing a drug to release. Evaporation and periodic sampling of the PBS results in the need to refill the tanks introducing error in the measurements. Also, the aromatic monomers used in this study were approximately \$400.00/g, and each well contained only 16µg of polymer. The microwell system obviates the need to sample PBS and uses small quantities of material lowering the cost to screen materials, which may have a major impact on new dendrimer and hyperbranched targeted release systems where only small amounts of material can be synthesized.

Biomaterial arrays were used by Langer and coworkers to screen human embryonic stem cell–material interactions [12]. Stem cells are cells that have not differentiated into a particular type of cell (e.g., heart cell, liver cell, or neuron). As a result, the possibility of using these cells to regenerate specific types of tissue may be possible if the cells are given the correct chemical cues to differentiate into the desired cell type. One possible method of inducing differentiation could be from different types of polymers. The authors used parallel synthesis to screen for these material–cell interactions. They coated glass slides with poly(hydroxyethyl methacrylate) (polyHEMA) and spotted a library of di- and triacrylates mixed with a photoinitiator. The resulting arrayed spots were then cured by exposing them to long-wavelength UV light. The microarrays were seeded with human embryonic stem cells to provide over 1700 cell–material interactions. The researchers found that nearly all the compositions supported cell attachment and growth as determined by staining and cell counts. However, when a soluble factor such as retinoic acid was added to the cell growth medium, the growth of cells was a stronger function of monomer type. Certain monomers allowed cell growth with and without the growth factor (Fig. 3.5). However, a couple of monomers prevented cell growth when retinoic acid was present, suggesting that cell growth and proliferation of human embryonic stem cells may be controlled with an appropriate choice of tissue-engineered construct.

In a follow-up to this work, a series of biodegradable polymer blends were arrayed on polyHEMA coated glass substrates [13]. The substrates were then seeded with human embryonic stem cells. The 24-member starting library of polymers resulted in a blend array of 3456 spots to screen cell–material interactions. Of note in this study was the lack of cell adhesion to neat polyethylene glycol (PEG) but restoration of cell attachment and growth when the PEG was blended with 30 wt% poly(*l*-lactide-co-glycolide) (*l*-PLGA) (70 : 30 l-lactide-co-glycolide). All other blends between PEG and PLGA with both stereoisomers of PLGA (*l* and *d*) resulted in no cell attachment. The authors believe that there may be phase separation between the PEG:*l*-PLGA blend resulting in regions rich in PEG and other regions rich in PLGA. As a consequence of phase separation, the *l*-PLGA regions must be large enough to provide area for cell attachment.

References 12 and 13 provide a relevant example of the power of combining synthesis or array technology with an existing immunohistochemistry characterization set. Each of the previous libraries was arrayed on standard glass slides that enabled the authors to use the standard fluorescence techniques of

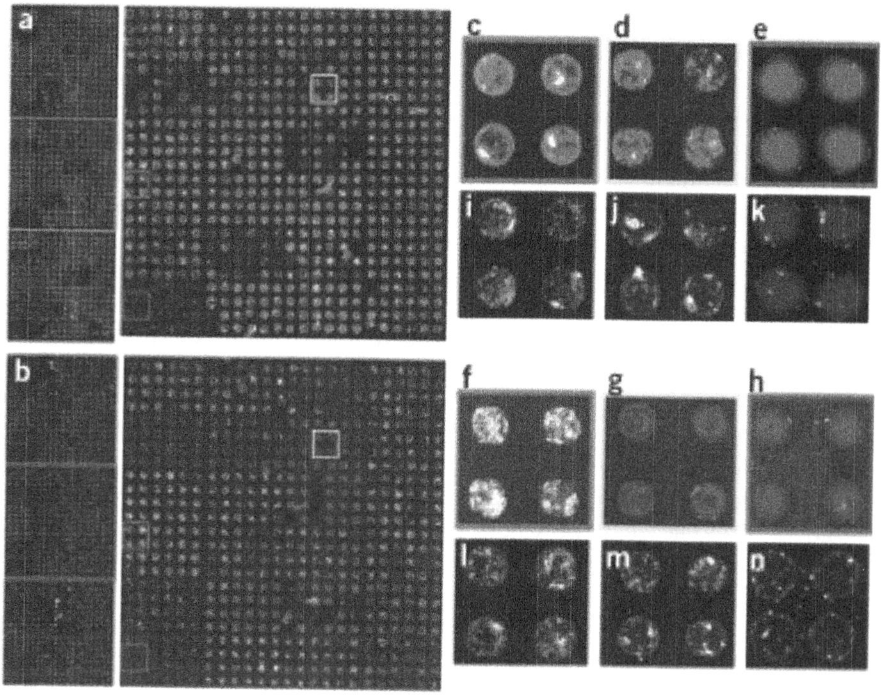

Figure 3.5. An example of human embryonic stem cells grown on a biomaterial array. Cells stained for cytokeratin 7 (green) and vimentin (red), grown in presence (a) and (b) absence of retinoic acid. The blue denotes polymer or cells lacking other signal by nuclear staining. Each spot represents a different polymer produced by UV crosslinking monomer formulations. (Images taken from Ref. 10.)

the cellular biologist. As methods are developed to further process the images of stained cells, more information and correlations will become available. These two studies emphasize the importance of polymer/material–cell interactions on not only the adhesion but also the growth and differentiation of stem cells. The basis of this work will provide guidelines for next-generation tissue scaffolds for organ and bone regeneration.

3.2.2. Polymers and Catalysts

Parallel techniques have been used as a method of screening functionalized polymers to find optimized compositions for use as catalytic media. For example, Menger and coworkers used parallel synthesis reactions to combinatorially functionalize amine containing polymers [poly(allylamine) and poly(ethylenimine)] with three to four carboxylic acids in the presence of 1-ethyl-3-(3-dimethylaminopropyl)-carbodiimide (EDC), trying to construct enzyme-like active sites that promote reductions of ketones to aldehydes [14]. Of the 8198 polymers synthesized, 92% were inactive in the reduction of benzoylformic acid to mandelic acid. However, they found that addition of a metal ion, such as magnesium, zinc, or copper, was necessary to get catalytic activity, and specific ratios of sidegroups were required for reductions to take place. The solvent choice also played a role in determining the activity of the functionalized polymer. The authors hypothesize that the solvent may play a role in the placement of sidegroups on the polymer backbone as well as order of functionality added. In a follow-up study, the authors simultaneously functionalized poly(acrylic anhydride) with three to four amines from a library of 11 amines for the discovery of a polymer catalyst for the dehydration of β-hydroxy ketone [15]. The authors found that enzymes still outperform the optimized combinatorially synthesized polymers but noted the polymers had a 24-h induction period before any catalytic activity was observed. They hypothesize that the polymers must "learn" to catalyze the ketone, which was explained in terms of a substrate-induced transformation into a catalytically active conformation.

Whitesides and coworkers generated libraries of combinatorially functionalized poly(acrylic acid) polymers to discover polyvalent inhibitors of the agglutination of erythrocytes induced by influenza virus A. The polymers were generated by modification of poly(acrylic anhydride) polymers with one or two amines and sialic acid, a functionality used by influenza virus A to adsorb and infect mammalian cells through the protein hemagglutinin (HA). The authors discovered that the ter-polymeric sialosides had the greatest binding affinity to HA and that the effect may be dominated by non-sialo sidegroups sterically attaching to the surface of the virus.

This set of papers reemphasizes the value of parallel synthesis not only for increased throughput but also the ability to vary parameters in a meaningful manner to systematically determine the effect of functional groups and interactions on, for instance, adsorption and catalytic activity. The power of combined foresight into monomer choice and throughput can lead to more focused libraries with the intent of discovering underlying phenomena, leading to the ability to rationally design systems for a given application. The papers reviewed in this section used combinatorial strategies to functionalize polymers that mimic biological polymers such as polypeptides and antibodies. Although precise synthetic control over placement of functionality along a polymer backbone does not yet exist, combinatorial chemistry can give one insight into the future of polymer synthesis. With more controlled synthetic techniques, combinatorial synthesis will become increasingly important because of functionality or the large number of monomers that will be incorporated into a given polymer. This will enable the precise control over molecular conformation or topology of the polymer that translate into the physical property of interest.

3.2.3. Functional Polymers and Dendrimers

Dendrimers are becoming increasingly important in catalysis, electronic materials, combinatorial synthesis supports, and biomaterials because of their high degree of peripheral functionality. Newkome and coworkers have developed a "graft to" approach to functionalize materials from dendrimers to cellulose. By preparing a series of isocyanate functional dendrons various moieties could be attached to hydroxy- and amine-containing polymers (Fig. 3.6). The authors used an extension of this technique to prepare combinatorially functionalized dendrimers and cellulose by adding more than one isocyanate dendron during the functionalization process. This allowed for deprotection of specific groups, letting the authors build up functionality and create dendrimer/hyperbranched hybrid molecules with properties between a linear polymer and a dendrimer [16–18]. The solubility of these materials can be tuned from water-soluble to organic solvent soluble by changing the ratio of dendrons and the functionality at the periphery of the dendrimer.

Similarly, Hawker and coworkers have used copper [3 + 2π] cycloadditions (a form of "click chemistry") of alkynes and azides to create libraries of dendrimers by functionalizing the periphery of an existing dendrimer core [19]. The click chemistry reaction is very specific and tolerant of delicate functional groups and can be performed in aqueous conditions. Another advantage of the cycloadditions was the near-quantitative yields and ease of creating azides (from a bromide and sodium azide). The hydroxyl dendrimers were functionalized with an alkyne by reaction with the anhydride of pentyonic acid.

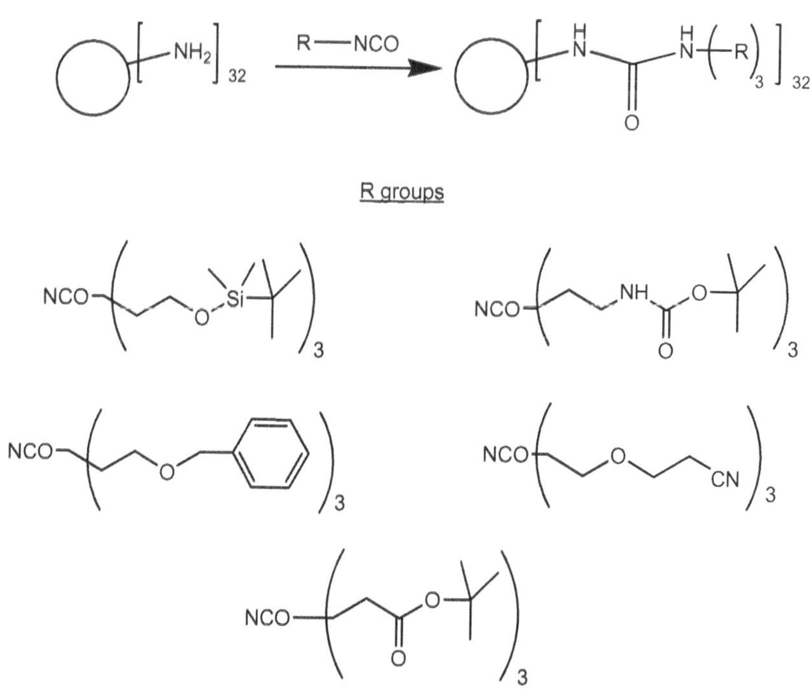

Figure 3.6. General scheme to functionalize alcohol and amine polymers.

Reaction of the alkyne polymer with one or more azides provided the functionality.

Schubert et al. have used combinatorial synthesis to explore unimolecular micelles and their sequestration of methyl orange [20,21]. Polyethylene glycol (PEG) 5-arm stars were used as macroinitiators for the ring-opening polymerization of ε-caprolactone with a tin initiator. The polymerization was performed using an automated ASW2000 robot under double inert atmosphere. A series of five different polymers were made ranging in number average relative molecular mass from 2,000 to 13,110 g/mol. The relative molecular masses were determined by GPC, proton NMR, and matrix-assisted laser desorption ionization time-of-flight mass spectrometry (MALDI-TOFMS). Copolymer composition was evaluated by a FTIR plate reader and the results agreed with increasing poly(ε-caprolactone) content. The host–guest capacity of each of the star copolymers and the dye methyl orange was determined in parallel by following the decay in the absorption maximum with ultraviolet–visible spectroscopy (UV-vis). The authors found the loading capacity was not a function of poly(ε-caprolactone) length but rather the polyethylene glycol core. The caprolactone shell only affected the solubility of the star in common solvents. In a follow-up study to this work, the authors screened the PEG-ε-caprolactone stars to capture 24 UV active dyes [21]; all except two of the dyes were taken into the star. The data were reportedly taken in less than 15 min, with the UV-vis taking about 40 s and the fluorescence measurements 2 min by use of 96 well microtiter plates.

A library of polyethylene oxide–polystyrene block (PEO-*b*-PS) copolymers was prepared by using by using a modular hydrogen bonding/supramolecular approach coined "macromolecular Lego's" [22]. A nitroxide initiator was prepared with a pendent terpyridine and used to polymerize polystryene. The resultant polystyrene chain contains a terpyridine at one end of the chain. Next, a monomethyl ether polyethylene oxide chain was functionalized with chloroterpyridine to provide an end-functional PEO. This polymer was allowed to react with ruthenium trichloride, and resulted in a PEO chain attached to a ruthenium(II) molecule. The PEO-Ru and PS-terpyridine were combined and a block copolymer formed. A 4 × 4 matrix of different PS and PEO relative molecular mass block copolymers were prepared and evaluated by atomic force microscopy (AFM) after spin coating on silicon oxide. Several different morphologies resulted depending on the different block lengths of the copolymers. This study is a good example of a modular approach to polymer synthesis that may lead to interesting new morphologies or polymer physics and engineering applications. The use of combinatorial synthesis and methods allowed Schubert and coworkers to quickly screen the parameter space and define interesting systems for more focused libraries.

Another advantage of parallel synthesis is the ability to determine reaction kinetics and screen for new polymerization catalysts in an efficient manner. Several groups have used parallel synthesis to explore conditions to perform such polymerizations as atom transfer radical polymerization (ATRP), reversible addition-fragmentation chain transfer (RAFT), xanthate polymerization [23,24], and living polymerizations [25]. The polymerization of methylmethacrylate by ATRP with five different metal salts, four initiators, and nine ligands, this library of 108 reactions revealed that Cu(I) systems were better controlled than were Fe(II)-mediated polymerizations under the same reactions conditions. It was also found the initiators *p*-toluenesulfonyl chloride and ethyl 2-bromoisobutyrate more effectively initiated polymerization over other bromo derivatives [26]. Moreover, the influence of the ligand was found to affect the final polydispersity index (PDI) more than the reaction kinetics.

Xanthates were shown to produce polymers and block copolymers with automated parallel synthesis by Destarac et al. [24] They showed the polymerizations were reproducible and matched conventionally polymerized samples. In comparison, the polymerization of acrylates and methacrylates with 2-cyano-2-butyl dithiobenzoate as a RAFT agent was optimized and it was found that after a reaction time of 10h the PDI of the final polymer increased dramatically due to loss of linearity in the first-order polymerization kinetics. A RAFT agent : initiator ratio of 4 : 1 was determined to be best with respect to PDI and relative molecular mass. In addition, living polymerizations have also been studied by parallel synthesis.

A series of papers explore the feasibility [27] and optimization with respect to temperature, activation energy [28,29], initiators, and copolymer composition [25] of the cationic ring-opening polymerization of 2-oxazolines. Furthermore, anionic polymerization of styrene, isoprene, and methylmethacrylate and their block copolymers was demonstrated within a parallel synthesis framework [30]. They also determined the apparent rate constant of the anionic polymerization of styrene in cyclohexane initiated by *sec*-butyllithium.

Parallel synthesis was used to explore the zinc-based catalysis of cyclohexene oxide with carbon dioxide to produce polycarbonates by Gruter et al. [31] This route provides an ecofriendly method of preparing polycarbonates when compared to the traditional phosgene route. One interesting result comes from this work, the activity of the catalyst was found to be much greater in the small reaction volumes of the parallel reactions than when scaled up. This was thought to be a result of mass transfer limitations caused by larger-scale equipment. This study reinforces the importance of scaling up from microscale reaction to lab- and plant-scale reactions. Furthermore, parallel polymerization screening reactions should always be verified for consistency with laboratory-scale reactions. In a another study, Woo and coworkers investigated the polymerization of norbornene by Ni(II), Pd(II), and Co(II) catalysts [32]. In general, Ni(II) complexes were more active than Pd(II) complexes and the Co(II) complexes were not active for any ligands tested.

The previous sections were devoted to the examples of the synthesis of polymeric discrete library through conventional methods. These methods consist of conducting reactions in microwells, vials, or reaction systems. The next section explores the philosophy, fabrication and use of microfluidics to prepare discrete libraries.

3.3. MICROFLUIDIC PLATFORMS FOR POLYMER LIBRARY PREPARATION AND CHARACTERIZATION

3.3.1. Introduction

Parallel reactors offer great advantages over conventional approaches to polymer synthesis. The speed and reduced error associated with automation should not be underestimated; however, the cost of such systems, particularly for more versatile equipment, is often prohibitive. Furthermore, the scale of the reactions, generally ≥5-mL reaction vessels, can be too large to screen many specialty materials such as biofunctional polymers, where starting material synthesis is carried out on a milligram scale. Alternate approaches to conventional parallel synthesis would be attractive for implementation in academia and small businesses, where capital equipment resources are limited.

Microfluidic technology, also known as "lab on a chip" and "micrototal analytical systems" (μTAS), has emerged as a popular platform to carry out biomedical research. Similar to the adoption of combinatorial chemistry for materials science, microfluidic devices are very appealing as environments to carry out chemical and materials research. The chemical industry is already benefiting from the many advantages that microreactor technology (of which microfluidics can be considered a subdiscipline) affords small-molecule reactions. Often the high surface : volume ratio and unique mixing environment produce improved heat and mass transfer, faster reactions, higher yields, and higher selectivity.

Microreactor technology, specifically microfluidics, has the potential to similarly benefit materials synthesis. There are unique demands on the reactor environment for materials handling that exceed the typical requirements of the pharmaceutical or chemical industries. These include high tolerance to aggressive organic solvents, high temperatures and pressures, and access to a range of length scales in manufacturing devices. Although all of these requirements have yet to be met in a single platform, considerable use has already been made of microscale environments, whether they consist of microchannels on chips, in microreactors or, potentially as combinations of both. This section seeks to briefly describe the variety of device fabrication routes available, to survey very recent work applying microfluidic (and microreactor) technology to polymer synthesis, and provide a few examples of high-throughput library design and fabrication made possible by microfluidic technology.

3.3.2. Microfluidic Device Fabrication

3.3.2.1. Glass and Silicon. Glass and silicon are commonly used materials for the fabrication of microchannels. Both materials, when properly patterned, treated, and sealed, provide resistance to organic solvents and high temperatures, a requirement for many chemical applications. Common photolithographic techniques were borrowed from the microelectronics industry to etch channel structures in these materials. This typically involves exposure of a thin layer of photoresist to radiation through a photomask, followed by chemical etching and bonding of the patterned surface to a flat surface for sealing.

Many of the first commercially available microfluidic devices were available in glass. The surface is amenable to functionalization with charged multilayers, facilitating electrokinetic flow, which is popular when working with aqueous solutions. Unfortunately, silicon does not have the optical transparency of glass, certain plastics, and elastomers, which is very appealing when integrating characterization tools. Some important work has been carried out in glass and silicon devices, including application of microfluidics to materials science for catalysis discovery and optimization [33–35].

3.3.2.2. Elastomers. Molding elastomeric materials against patterned surfaces to produce channel structures is one of the most popular routes to preparing microfluidic devices. Often referred to as "soft lithography" or "rapid prototyping" [36], this process produces patterns that can then be sealed against glass or elastomeric surfaces quickly and easily to produce a working device. Most applications of this approach use polydimethylsiloxane (PDMS) as the elastomer. PDMS can be formulated to exhibit a range of moduli; however, the most commonly used commercial product (Sylgard 184 from Dow Corning) has a modulus of about 2–5 MPa, making it easy to work with (releasing from master patterns, coring connections, etc.).

The first applications of microfluidics to materials were carried out inside PDMS devices. For example, the synthesis of colloidal silica using segmented plug flow in a PDMS channel produced tunable particle sizes and size distributions by variation of velocity and residence time of the plugs [37]. CdS and CdSe nanoparticles have also been prepared [38]. More work with PDMS devices for the fabrication of polymer particles is discussed later in the chapter.

The main disadvantage to working with PDMS devices for polymer applications is its poor compatibility with most organic solvents. The development of solvent-resistant elastomers with mechanical properties similar to PDMS turns out to be fairly challenging, but was recently accomplished with the development of a curable perfluoropolyether [39], which has similar mechanical performance to PDMS with excellent chemical stability [40].

3.3.2.3. Plastics. Patterning of channels in plastics is appealing because of the reduced materials cost and increased durability of many plastic materials [41,42]. In many cases, plastics also maintain the transparency of glass and many elastomeric materials. There are several approaches to patterning plastics, including injection molding [43], hot embossing (or imprinting) [44,45], and laser ablation [46,47]. Injection molding and hot embossing are hampered by the same limitation as are the elastomeric materials discussed above. The material must be molded against a master, which can be expensive and time-consuming. Furthermore, both techniques are limited to materials that can be processed at high temperatures. For high-volume manufacturing, however, this remains an attractive process to fabricate microfluidic devices.

A relatively new approach combines technology borrowed from the microelectronics industry with patterning of polymeric materials into microfluidic channels [48]. A variety of materials have been used in this process, including acrylics [48], urethanes [36], and epoxies [49]. More recently, a similar approach was used to construct devices in thiolene networks [50,51]. Three major advantages are obtained when using thiolene to make devices: (1), thiolene devices have improved solvent resistance to aromatic and aliphatic solvents; (2) thiolene materials cure by a frontal photopolymerization, meaning that a sharp solid–liquid boundary originating at the exposure surface grows into the material as a function of UV dose; and (3) this enables variable height patterning of features in the same device using either gray-scale masking or sequential exposures of different UV doses with different mask patterns [52].

Photocurable plastics can also accelerate the rapid prototyping process of preparing elastomeric microfluidic devices. Use of a negative mask design and a single plate (replacing one plate with a release layer) in contact with the polymerizing liquid can produce a master against which elastomers can be molded and released [50,53]. This eliminates the need for expensive silicon masters. All fabrication can be performed in a typical lab environment with inexpensive equipment and materials. Figure 3.7 shows a cross section of the fabrication process for either a closed device or a prototyping master.

In summary, polymeric materials have become the most appealing medium for constructing microfluidic devices, and polymer science has advanced the versatility and function of microfluidic technology in many ways beyond those mentioned here [54–56]. As chip design and fabrication methods become more mature and readily accessible, scientists will expand their applications. Fields such as polymeric materials synthesis and polymer physics should benefit

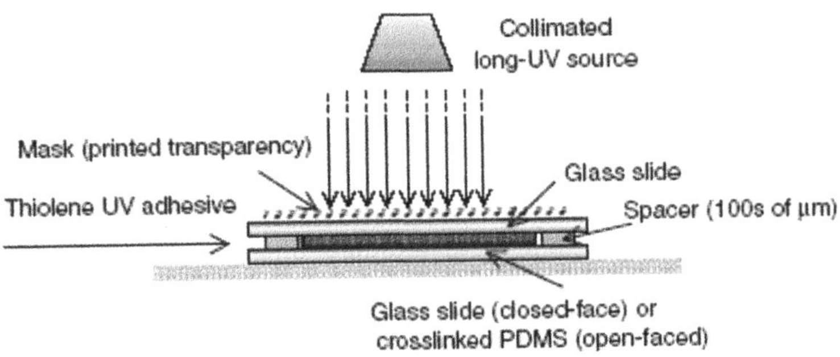

Figure 3.7. Cross-sectional schematic of the fabrication process for plastic microfluidic devices. (Image reprinted with permission from Ref. 50.)

greatly from this expanding technology. Synthetic methods have been the first significant application area, with a number of groups demonstrating the potential power of microreactors and microfluidic synthesis for polymeric materials.

3.3.3. Polymer Synthesis

3.3.3.1. Droplets and Particles. The manipulation and control of two-phase flows in microchannels has developed into a rich field of study over the last several years. Droplet or plug formation, generally using one of two channel designs (referred to as "T junctions" [57,58] and "flow focusing" [59,60]), has become fairly common. Application of droplets as individual microreactors has been demonstrated for real-time monitoring of reaction kinetics [61], protein crystallization [62], and controlled colloidal assembly [63].

The first polymerizations facilitated by microfluidic devices were carried out in droplets suspended in a continuous phase. Using a silicon microchannel, monodisperse emulsions with monomer droplet sizes of 50 μm or more were produced in a controlled manner [64,65]. Beebe and coworkers later demonstrated UV polymerization of fibers and tubes inside a microchannel [66]. Droplet formation combined with UV polymerization inside a microchannel was used to produce monodisperse polymer particles [67,68]. Spherical, discoidal, and pellet particle shapes could be obtained depending on the size of the droplet relative to the cross-sectional area and aspect ratio of the channel. All of these polymerizations were carried out in PDMS devices.

Another innovative approach to manipulating microdroplet reactors has been demonstrated by Velev and coworkers [69,70]. Aqueous and aliphatic droplets were suspended on a layer of fluorinated oils and dielectrophoretically controlled by voltage switches across a series of electrodes to induce motion across the surface. Unique anisotropic particles were also prepared using this technique.

In general, these methods have been used as manufacturing tools to make ensembles of uniform particles. Typically, it is the polymer particle itself that is the end goal in these reactions, which are carried out with cross-linking monomers and/or with uncontrolled radical reactions to produce high-relative-molecular-mass solid particles. In fact, one key advantage of microfluidics from this perspective is the ability to minimize particle size distribution, which is a major challenge in industrial production.

3.3.3.2. Continuous-Phase Solution Polymerization. Polymerization within the continuous phase of a microchannel remains a challenging, but potentially rewarding, environment in which to carry out polymer synthesis. Two of the features of microreactors, fast mixing and precise residence (i.e., reaction) time control, in addition to excellent heat transfer, have particular benefits to solution phase and low molecular-weight polymerization. Removal of the second phase eliminates a separation step for product isolation, and more closely approximates the analogous macroscale continuous processes, for which there are already models for real-time reaction monitoring.

Many of the well-defined, valuable polymeric additives in soft materials are of low relative molecular mass (<25,000 g/mol). Often chain sequence in block and statistical copolymers directly influences material properties, in both complex solutions and bulk/solid phases. Recent advances in controlled and living polymer synthesis, combined with intense research in biomaterials and nanotechnology, have enabled an expansion in the design parameters of these materials and established a need for precision control in macromolecular architecture.

One of the keys to a controlled polymerization mechanism is the establishment of a dynamic equilibrium between active and dormant chain ends to suppress the concentration of active species and minimize side reactions. This tends to slow down the overall rate of polymerization significantly and often requires the use of transition metals or undesirable organic additives. Even if one were able to accelerate the rate of polymerization, initiation would have to remain sufficiently fast relative to propagation.

Microreactors were recently used to test this strategy with some success for both cationic (referred to as the "cation pool" method) [71] and radical polymerizations [72], where improved control was attributed to both rapid mixing (for fast initiation) and improved heat transfer. In both demonstrations, there was no deactivator present in solution, meaning that the reactions were significantly faster than the typical controlled reaction, while obtaining molecular weight distributions intermediate between conventional and controlled reactions for most of the monomers studied.

The only continuous polymerizations carried out in a microfluidic device used atom transfer radical polymerization (ATRP) to control the active species concentration. Referred to as "controlled radical polymerization on a chip" (CRP chip), this approach slowed the rate of polymerization to the batch reactor rates [73]. The goal of this work was slightly different (see discussion below); however, similar improvements in the rate of polymerization were reported in the case where the conditions used did not produce optimal relative molecular mass distribution control in the batch reactions [74]. These results are surprising as they suggest suppression of side reactions by the slow reaction rate, beyond that afforded by improved heat transfer (1–2 h for high conversions with low-relative-molecular-mass targets). It should also be noted that these reactions required flow rates far below those typically used in microfluidic devices to accommodate the comparatively long reaction times.

The basic reactor designs for these types of polymerization are shown in Figure 3.8. With only a few demonstrations, these results suggest there is a wealth of opportunity in the area of understanding polymerizations confined within microchannels.

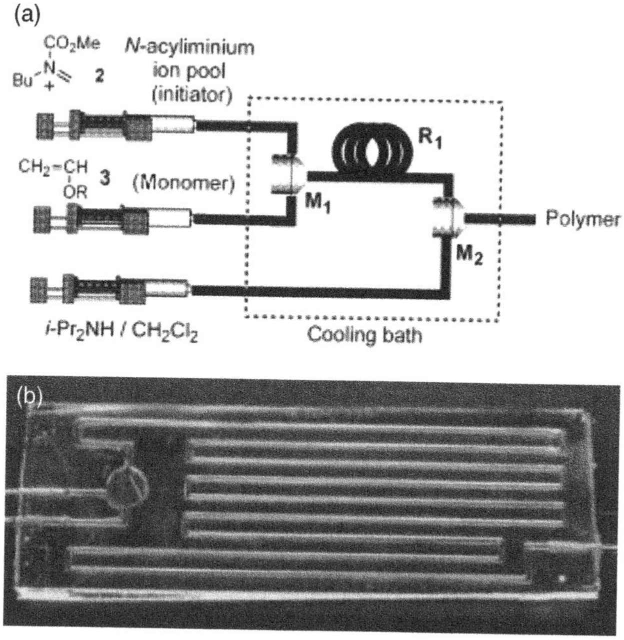

Figure 3.8. (a) Schematic representation of microreactor design for the "cation pool" method of controlled cationic polymerization. (Reprinted with permission from Ref. 71.) (b) Image of a microchip used for controlled radical polymerization on a chip. Monomer/initiator and monomer/catalyst solutions flow in from the left, are mixed, and flow through the microchannel. The reaction is quenched at the outlet. Flow rate controls residence time, which correlates to reaction time and monomer conversion. (CRP chip; reprinted with permission from Ref. 73.)

3.3.4. Library Design and Fabrication

To this point, all of the applications discussed have involved the use of microreactors and microfluidics for continuous production of uniform samples. Another application area for this technology is combinatorial and high-throughput library synthesis. This section outlines basic strategies and three ways to implement microfluidic technology to prepare polymer libraries.

Several reviews describe the use of microchannels and microfluidics to carry out combinatorial chemistry on small organic molecule libraries [75,76] or to monitor catalytic onstream function of heterogeneous catalysts [77,78]. These techniques can be strictly high-throughput; for example, rapid variations in multiple input flow rates, or flow switching, can produce a large number of reaction conditions by exploiting the spatial and temporal control of microchannels. Similarly, rapid in situ analysis can be performed because of the relatively high yields of pure products produced compared to those using conventional reactor methods. Microreactors can also be designed for parallel reactions and monitoring by either stacking multiple chips or manufacturing chips with multiple channels for classical combinatorial experiments.

3.3.4.1. Droplets. Changing relative flow rates to change the composition of droplets as they are forming has been demonstrated to screen protein crystallization, by controlling the relative concentrations of polymer, protein, and salt in a sequence of aqueous droplets produced in a T junction [62]. A similar strategy was used to control stoichiometry in a bromination reaction by changing the relative flowrates of bromine and styrene forming organic droplets in a flow-focusing device [51]. This device structure has been used to create an array of droplets with changing monomer compositions. Figure 3.9 contains images of arrays prepared with organic dyes in the monomer solutions to visually represent the changing monomer composition in the droplets. Fiberoptic Raman spectroscopy was used to monitor composition and monomer conversion of the droplets. Polymerization was initiated with UV light from a liquid spot probe to form highly crosslinked glassy particles [79]; styrenic and methacrylic particles were prepared.

When reaction timescales are long (greater than a few minutes), maintaining a steady flowrate without clogging or coalescing droplets becomes more difficult. For fast reactions, droplets can be monitored under flow using detectors at fixed points or imaging tools. Stationary arrays can be constructed in the device and monitored using translation stages, for single-point monitoring (high throughput), and/or imaging microscopes (combinatorial). Because the

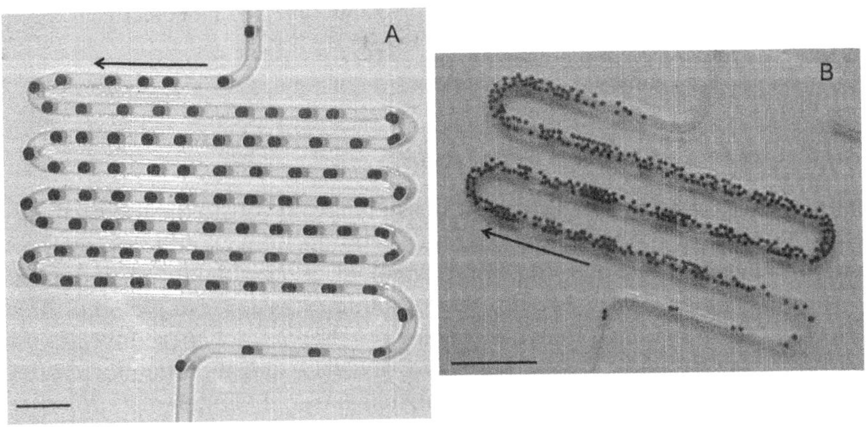

Figure 3.9. Arrays of droplets constructed using flow focusing inside a microfluidic device for high-throughput screening. Red and blue dyes in each methacrylate monomer solution enable visual representation of the composition gradient in the array. (a) Large droplets for a single microreactor per experiment; (b) smaller droplets allow averaging over multiple microreactors for statistics on behavior such as droplet size as a function of composition and extent of reaction (i.e., monomer conversion). Arrows indicate direction of flow. Scale bars represent 5 mm. (Images from Ref. 79.)

devices are disposable and can be redesigned and fabricated quickly, the technique is very flexible.

3.3.4.2. Continuous-Phase Polymerizations. Less high-throughput work has been carried out in the area of continuous-phase polymerizations. One approach involved the use of multiple fluid inlets to control the relative concentrations of monomer, macroinitiator, and catalyst in the CRP chip [74]. Varying either the flowrate or concentrations produced a series of block copolymers with different lengths of the second block. Sample volumes ranged from a few microliters to milliliters depending on the collection time at each set of conditions. While the reactions do not proceed faster in this experiment, many polymer products can be prepared from a single batch of three solutions, and by incorporating measurement tools into the devices, property screening on a broad range of polymers can be carried out without using large quantities of starting materials. Small-scale measurement tools become necessary when starting materials are rare or difficult to produce. Continuous phase polymerizations on a chip also have the potential to produce a lab-on-a-chip analog to a new reaction monitoring method referred to as "automatic continuous online monitoring of polymerization reactions" (ACOMP) [80,81].

3.3.4.3. Polymer Brushes. Covalently tethered polymer chains on surfaces, often called "polymer brushes," are another area of wide interest [82–84]. Polymer brushes have many of the same benefits as self-assembled monolayers for controlling surface chemistry for separations [85], biointerfaces [86,87], and other processes. Brushes, however, have the advantage of being significantly more stable, in particular when grafted to silicon, as well as providing some control of surface mechanical properties [88].

Surfaces designed with gradual variation in brush properties were demonstrated previously [89–91]. These gradient surfaces can be used to rapidly map interactions between polymer brushes and other materials as a function of one or two variables. Using existing techniques, however, the gradients are restricted to a single axis per substrate (Fig. 3.10a).

In order to produce single surfaces that express multiple variables, spatial control of the surface chemistry is critical. Microchannels provide the confinement necessary to achieve control over brush-forming reactions as they proceed. A recent demonstration of microchannel-confined surface-initiated polymerization (μSIP; Fig. 3.10b) allowed the construction of gradients whose properties were determined by the removable elastomeric channel that confined the monomer/catalyst solution on the surface and the flow rate of the solution [92]. Furthermore, the channels could be repeatedly overlaid to construct layered patterns on the surface, and the fluid composition itself could be controlled spatially within a microchannel, constructing a continuous gradient in the polymerization conditions. The latter method was used to produce a gradient in statistical copolymer composition [93]. This variable in brush architecture is difficult to produce using other methods and enables the mixed

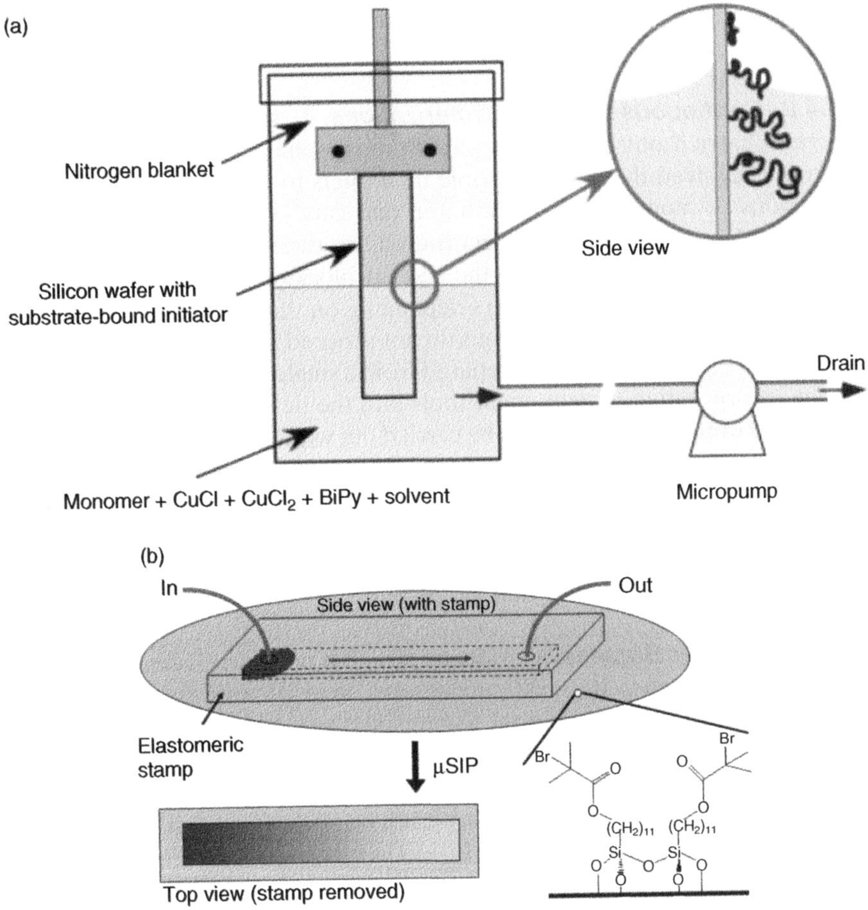

Figure 3.10. Schematic representation of the two basic approaches to producing gradients of grafted polymer chains. [(a) Reproduced with permission from Ref. 90; (b) reproduced with permission from Ref. 92.]

expression of chemical species at the interfaces by suppressing phase segregation [94]. So far, the application of μSIP has been restricted to demonstration of the fabrication capabilities, such as the types of gradients that can be produced.

3.3.5. Future Directions

Fast, inexpensive library synthesis strategies will increase in importance as molecular architectures become more complex, starting materials become more expensive and difficult to make in large quantities, and as more discovery and innovation occurs in academia and small businesses. To make these tools relevant to material discovery needs, the corresponding measurement needs will have to be met. Biologically relevant materials and nanostructured materials represent two attractive application areas. There are two basic strategies to carry out this integration.

On-chip fabrication, processing, and measurement elements with versatile designs can create a toolbox from which to tailor multifunctional chips. Measurements to characterize complex polymeric solutions and fluids on a chip remain a relatively unexplored area. This includes rheology as well as structural characterization, such as light and X-ray scattering. As an alternative, chip-scale measurements that do not require microfluidic channels, but do require sample volumes commensurate with those prepared in microfluidic devices, also play a critical role. However, several promising new measurements of mechanical [95] and adhesive [96] properties possess these capabilities.

3.4. SUMMARY

This chapter provides a thorough review of the literature pertaining to the synthesis of conventional parallel and high-throughput microfluidic polymer discrete libraries. A discussion of the application of high-throughput methods to such fields as biomaterials, polymer catalysis, functional polymers and dendrimers, and microfluidics yielded a few salient points regarding these methods:

1. Discrete libraries are a powerful tool for systematic screening of a parameter space for desired properties; however, optimization is a process and large libraries should be focused into smaller libraries.
2. Parallel methods will have a large impact in biomaterials due to the current infrastructure already in place from biological characterization methods.
3. A high-throughput study is only as good as the techniques used to characterize the system.
4. Lab-on-a-chip methods will become increasingly more important because of the low-cost fabrication methods to prepare the chips and the ability to integrate not only synthesis but also onboard separations and characterization.

REFERENCES

1. Schmatloch, S., Meier, M. A. R., and Schubert, U. S., Instrumentation for combinatorial and high-throughput polymer research: A short overview, *Macromol. Rapid Commun.* **24**(1):33–46 (2003).

2. Dar, Y. L., High-throughput experimentation: A powerful enabling technology for the chemicals and materials industry, *Macromol. Rapid Commun.* **25**(1):34–47 (2004).
3. Brocchini, S., James, K., Tangpasuthadol, V., and Kohn, J., A combinatorial approach for polymer design, *J. Am. Chem. Soc.* **119**(19):4553–4554 (1997).
4. Brocchini, S., James, K., Tangpasuthadol, V., and Kohn, J., Structure-property correlations in a combinatorial library of degradable biomaterials, *J. Biomed. Mater. Res.* **42**(1):66–75 (1998).
5. Rickerby, J., Prabhakar, R., Patel, A., Knowles, J., and Brocchini, S., A biomedical library of serinol-derived polyesters, *J. Control. Release* **101**(1–3):21–34 (2005).
6. Akinc, A., Lynn, D. M., Anderson, D. G., and Langer, R., Parallel synthesis and biophysical characterization of a degradable polymer library for gene delivery, *J. Am. Chem. Soc.* **125**(18):5316–5323 (2003).
7. Lynn, D. M., Anderson, D. G., Putnam, D., and Langer, R., Accelerated discovery of synthetic transfection vectors: Parallel synthesis and screening of degradable polymer library, *J. Am. Chem. Soc.* **123**(33):8155–8156 (2001).
8. Anderson, D. G., Lynn, D. M., and Langer, R., Semi-automated synthesis and screening of a large library of degradable cationic polymers for gene delivery, *Angew. Chem. Int. Ed.* **42**(27):3153–3158 (2003).
9. Anderson, D. G., Akinc, A., Hossain, N., and Langer, R., Structure/property studies of polymeric gene delivery using a library of poly(beta-amino esters), *Mol. Ther.* **11**(3):426–434 (2005).
10. Anderson, D. G., Peng, W. D., Akinc, A., Hossain, N., Kohn, A., Padera, R., et al., A polymer library approach to suicide gene therapy for cancer, *Proc. Natl. Acad. Sci. USA* **101**(45):16028–16033 (2004).
11. Vogel, B. M., Cabral, J. T., Eidelman, N., Narasimhan, B., and Mallapragada, S. K., Parallel synthesis and high throughput dissolution testing of biodegradable polyanhydride copolymers, *J. Combin. Chem.* **7**(6):921–928 (2005).
12. Anderson, D. G., Levenberg, S., and Langer, R., Nanoliter-scale synthesis of arrayed biomaterials and application to human embryonic stem cells, *Nature Biotechnol.* **22**(7):863–866 (2004).
13. Anderson, D. G., Putnam, D., Lavik, E. B., Mahmood, T. A., and Langer, R., Biomaterial microarrays: rapid, microscale screening of polymer-cell interaction, *Biomaterials* **26**(23):4892–4897 (2005).
14. Menger, F. M., West, C. A., and Ding, J., A combinatorially developed reducing agent, *Chem. Commun.* **1997**(6):633–634 (1997).
15. Menger, F. M., Ding, J., and Barragan, V., Combinatorial catalysis of an elimination reaction, *J. Org. Chem.* **63**(22):7578–7579 (1998).
16. Newkome, G. R., Suprasupermolecular chemistry: The chemistry within the dendrimer, *Pure Appl. Chem.* **70**(12):2337–2343 (1998).
17. Newkome, G. R., Childs, B. J., Rourk, M. J., Baker, G. R., and Moorefield, C. N., Dendrimer construction and macromolecular property modification via combinatorial methods, *Biotechnol. Bioeng.* **61**(4):243–253 (1999).
18. Newkome, G. R., Weis, C. D., Moorefield, C. N., Baker, G. R., Childs, B. J., and Epperson, J., Isocyanate-based dendritic building blocks: Combinatorial tier

construction and macromolecular-property modification, *Angew. Chem. Int. Ed.* **37**(3):307–310 (1998).

19. Malkoch, M., Schleicher, K., Drockenmuller, E., Hawker, C. J., Russell, T. P., Wu, P., et al., Structurally diverse dendritic libraries: A highly efficient functionalization approach using Click chemistry, *Macromolecules* **38**(9):3663–3678 (2005).
20. Meier, M. A. R., Gohy, J. F., Fustin, C. A., and Schubert, U. S., Combinatorial synthesis of star-shaped block copolymers: Host-guest chemistry of unimolecular reversed micelles, *J. Am. Chem. Soc.* **126**(37):11517–11521 (2004).
21. Meier, M. A. R. and Schubert, U. S., Combinatorial evaluation of the host-guest chemistry of star-shaped block copolymers, *J. Combin. Chem.* **7**(3):356–359 (2005).
22. Lohmeijer, B. G. G., Wouters, D., Yin, Z. H., and Schubert, U. S., Block copolymer libraries: Modular versatility of the macromolecular Lego (R) system, *Chem. Commun.* **2004**(24):2886–2887 (2004).
23. Fiten, M. W. M., Paulus, R. M., and Schubert, U. S., Systematic parallel investigation of RAFT polymerizations for eight different (meth)acrylates: A basis for the designed synthesis of block and random copolymers, *J. Polym. Sci. (Pt. A—Polym. Chem.)* **43**(17):3831–3839 (2005).
24. Chapon, P., Mignaud, C., Lizarraga, G., and Destarac, M., Automated parallel synthesis of MADIX (Co)polymers, *Macromol. Rapid Commun.* **24**(1):87–91 (2003).
25. Hoogenboom, R., Fijten, M. W. M., and Schubert, U. S., Parallel kinetic investigation of 2-oxazoline polymerizations with different initiators as basis for designed copolymer synthesis, *J. Polym. Sci. (Pt. A—Polym. Chem.)* **42**(8):1830–1840 (2004).
26. Zhang, H. Q., Marin, V., Fijten, M. W. M., and Schubert, U. S., High-throughput experimentation in atom transfer radical polymerization: A general approach toward a directed design and understanding of optimal catalytic systems, *J. Polym. Sci. (Pt. A—Polym. Chem.)* **42**(8):1876–1885 (2004).
27. Hoogenboom, R., Fijten, M. W. M., Meier, M. A. R., and Schubert, U. S., Living cationic polymerizations utilizing an automated synthesizer: High-throughput synthesis of polyoxazolines, *Macromol. Rapid Commun.* **24**(1):92–97 (2003).
28. Hoogenboom, R., Fijten, M. W. M., Brandli, C., Schroer, J., and Schubert, U. S., Automated parallel temperature optimization and determination of activation energy for the living cationic polymerization of 2-ethyl-2-oxazoline, *Macromol. Rapid Commun.* **24**(1):98–103 (2003).
29. Hoogenboom, R., Fijten, M. W. M., and Schubert, U. S., The effect of temperature on the living cationic polymerization of 2-phenyl-2-oxazoline explored utilizing an automated synthesizer, *Macromol. Rapid Commun.* **25**(1):339–343 (2004).
30. Guerrero-Sanchez, C., Abeln, C., and Schubert, U. S., Automated parallel anionic polymerizations: Enhancing the possibilities of a widely used technique in polymer synthesis, *J. Polym. Sci. (Pt. A—Polym. Chem.)* **43**(18):4151–4160 (2005).
31. van Meerendonk, W. J., Duchateau, R., Koning, C. E., and Gruter, G. J. M., High-throughput automated parallel evaluation of zinc-based catalysts for the copolymerization of CHO and CO2 to polycarbonates, *Macromol. Rapid Commun.* **25**(1):382–386 (2004).

66. Jeong, W., Kim, J., Kim, S., Lee, S., Mensing, G., and Beebe, D. J., Hydrodynamic microfabrication via "on the fly" photopolymerization of microscale fibers and tubes, *Lab Chip* **4**(6):576–580 (2004).
67. Xu, S. Q., Nie, Z. H., Seo, M., Lewis, P., Kumacheva, E., Stone, H. A., et al., Generation of monodisperse particles by using microfluidics: Control over size, shape, and composition, *Angew. Chem. Int. Ed. Engl.* **44**(5):724–728 (2005).
68. Dendukuri, D., Tsoi, K., Hatton, T. A., and Doyle, P. S., Controlled synthesis of nonspherical microparticles using microfluidics, *Langmuir* **21**(6):2113–2116 (2005).
69. Velev, O. D., Prevo, B. G., and Bhatt, K. H., On-chip manipulation of free droplets, *Nature* **426**(6966):515–516 (2003).
70. Millman, J. R., Bhatt, K. H., Prevo, B. G., and Velev, O. D., Anisotropic particle synthesis in dielectrophoretically controlled microdroplet reactors, *Nature Mater.* **4**(1):98–102 (2005).
71. Nagaki, A., Kawamura, K., Suga, S., Ando, T., Sawamoto, M., and Yoshida, J., Cation pool-initiated controlled/living polymerization using microsystems, *J. Am. Chem. Soc.* **126**(45):14702–14703 (2004).
72. Iwasaki, T. and Yoshida, J., Free radical polymerization in microreactors. Significant improvement in molecular weight distribution control, *Macromolecules* **38**(4):1159–1163 (2005).
73. Wu, T., Mei, Y., Cabral, J. T., Xu, C., and Beers, K. L., A new synthetic method for controlled polymerization using a microfluidic system, *J. Am. Chem. Soc.* **126**(32):9880–9881 (2004).
74. Wu, T., Mei, Y., Xu, C., Byrd, H. C. M., and Beers, K. L., Block copolymer PEO-b-PHPMA synthesis using controlled radical polymerization on a chip, *Macromol. Rapid Commun.* **26**(13):1037–1042 (2005).
75. Suga, S., Okajima, M., Fujiwara, K., and Yoshida, J., Electrochemical combinatorial organic synthesis using microflow systems, *QSAR Combin. Sci.* **24**:728–741 (2005).
76. Watts, P. and Haswell, S. J., Microfluidic combinatorial chemistry, *Curr. Opin. Chem. Biol.* **7**:380–387 (2003).
77. Senkan, S., Combintatorial heterogeneous catalysis—a new path in an old field, *Angew. Chem. Int. Ed. Engl.* **40**:312–329 (2001).
78. Senkan, S. M. and Ozturk, S., Discovery and optimization of heterogeneous catalysts by using combinatorial chemistry, *Angew. Chem. Int. Ed. Engl.* **38**:791–795 (1999).
79. Barnes, S. E., Cygan, Z. T., Yates, J. K., Beers, K. L., and Amis, E. J., Raman spectroscopic monitoring of droplet polymerization in a microfluidic device. *Analyst* **131**:1027–1033 (2006).
80. Giz, A., Catalgil-Giz, H., Alb, A., Brousseau, J. L., and Reed, W. F., Kinetics and mechanisms of acrylamide polymerization from absolute, online monitoring of polymerization reaction, *Macromolecules* **34**(5):1180–1191 (2001).
81. Florenzano, F. H., Strelitzki, R., and Reed, W. F., Absolute, on-line monitoring of molar mass during polymerization reactions, *Macromolecules* **31**(21):7226–7238 (1998).

82. Edmondson, S., Osborne, V. L., and Huck, W. T. S., Polymer brushes via surface-initiated polymerizations, *Chem. Soc. Rev.* **33**(1):14–22 (2004).

83. Zhao, B. and Brittain, W. J., Polymer brushes: Surface-immobilized macromolecules, *Progress Polym. Sci.* **25**(5):677–710 (2000).

84. Leger, L., Raphael, E., and Hervet, H., Surface-anchored polymer chains: Their role in adhesion and friction, in *Polymers in Confined Environments*, pp. 185–225, S. Granick and K. Binder (eds.), Springer, New York, NY, 1999.

85. Miller, M. D., Baker, G. L., and Bruening, M. L., Polymer-brush stationary phases for open-tubular capillary electrochromatography, *J. Chromatogr. A* **1044**(1–2):323–330 (2004).

86. Mei, Y., Wu, T., Xu, C., Langenbach, K. J., Elliott, J. T., Vogt, B. D., et al., Tuning cell adhesion on gradient poly(2-hydroxyethyl methacrylate)-grafted surfaces, *Langmuir* **21**(26):12309–12314 (2005).

87. Bhat, R. R., Chaney, B. N., Rowley, J., Liebmann-Vinson, A., and Genzer, J., Tailoring cell adhesion using surface-grafted polymer gradient assemblies, *Adv. Mater.* **17**(23):2802 (2005).

88. Lemieux, M., Usov, D., Minko, S., Stamm, M., Shulha, H., and Tsukruk, V. V., Reorganization of binary polymer brushes: Reversible switching of surface microstructures and nanomechanical properties, *Macromolecules* **36**(19):7244–7255 (2003).

89. Bhat, R. R., Tomlinson, M. R., and Genzer, J., Orthogonal surface-grafted polymer gradients: A versatile combinatorial platform, *J. Polym. Sci. (Pt. B—Polym. Phys.)* **43**(23):3384–3394 (2005).

90. Tomlinson, M. R. and Genzer, J., Formation of grafted macromolecular assemblies with a gradual variation of molecular weight on solid surfaces, *Macromolecules* **36**(10):3449–3451 (2003).

91. Wu, T., Tomlinson, M., Efimenko, K., and Genzer, J., A combinatorial approach to surface anchored polymers, *J. Mater. Sci.* **38**(22):4471–4477 (2003).

92. Xu, C., Wu, T., Drain, C. M., Batteas, J. D., and Beers, K. L., Microchannel confined surface-initiated polymerization, *Macromolecules* **38**(1):6–8 (2005).

93. Xu, C., Barnes, S. E., Wu, T., Fischer, D. A., DeLongchamp, D. M., Batteas, J. D., et al., Solution and surface gradients via microfluidic confinement: Fabrication of a statistical copolymer brush composition gradient, *Adv. Mater.* **18**:1427–1430 (2006).

94. Ryu, D. Y., Shin, K., Drockenmuller, E., Hawker, C. J., and Russell, T. P., *Science* **308**:236 (2005).

95. Stafford, C. M., Harrison, C., Beers, K. L., Karim, A., Amis, E. J., Vanlandingham, M. R., et al., A buckling-based metrology for measuring the elastic moduli of polymeric thin films, *Nature Mater.* **3**(8):545–550 (2004).

96. Crosby, A. J., Karim, A., and Amis, E. J., Combinatorial investigations of interfacial failure, *J. Polym. Sci. (Pt. B—Polym. Phys.)* **41**(9):883–891 (2003).

CHAPTER 4

Strategies in the Use of Atomic Force Microscopy as a Multiplexed Readout Tool of Chip-Scale Protein Motifs

JEREMY R. KENSETH, KAREN M. KWARTA, and JEREMY D. DRISKELL

Institute for Combinatorial Discovery
Ames Laboratory—USDOE
Department of Chemistry
Iowa State University
Ames, Iowa

MARC D. PORTER

Department of Chemistry and Biochemistry
Center for Combinatorial Science at The Biodesign Institute
Arizona State University
Tempe, Arizona

JOHN D. NEILL and JULIA F. RIDPATH

Virus and Prion Diseases of Livestock Unit
National Animal Disease Center
United States Department of Agriculture (USDA)
Ames, Iowa

4.1. INTRODUCTION

Combinatorial science is redefining laboratory convention by employing massive sample libraries and high-throughput testing and readout strategies in order to transcend the one-experiment-at-a-time modality of the traditional scientific method [1–6]. These developments span areas from drug discovery and materials science to early disease detection and homeland security. Concepts based on combinatorial science are also being applied to advance the fundamental understanding of several chemical and biochemical processes.

The majority of today's screening processes utilize optical spectroscopy, mass spectrometry, and electrochemical techniques to identify targets and quantify their performance. In the case of early disease detection and home-

Combinatorial Materials Science, Edited by Balaji Narasimhan, Surya K. Mallapragada, and Marc D. Porter

labeling proteins on cell surfaces with electron microscopy [39–41]; in some instances, different-sized particles have been employed to selectively label different cell receptors [42].

By using different sized labels, it is possible to simultaneously determine the presence of multiple analytes in a sample, provided that the imaging technique can adequately resolve the differences in label size. Several laboratories [42–51] have explored the microscopic characterization capabilities of AFM for the potential development of high-throughput, miniaturized immunoassays. We recently demonstrated the concept of a height-based AFM immunoassay, in which the changes in topography that result from the binding of an antibody to immobilized micrometer-sized array elements of antigen were utilized [52,53]. The motivation for these investigations arises from the desire to improve detection limits, reduce sample preparation, and decrease the amount of sample required for analysis [54]. Since AFM is capable of imaging surfaces with nanometer-scale resolution there are a wide range of size-based analyses that can be developed [55,56].

Both scanning tunneling microscopy [47,57] and AFM [43,44,58–66] have been employed to image gold-conjugated antibodies on surfaces. Gold conjugates have been used to verify immobilization of antibodies to examine relative binding efficiencies of different antibody immobilization schemes [58–60], to localize cellular surface proteins [61–64,66], and to map domains of DNA [65]. Past reports have also demonstrated that it is possible to conduct highly sensitive, sandwich immunoassays by imaging gold-conjugated antibodies with AFM [44]. Moreover, a detection limit of 0.02 pg/mL for thyroid-stimulating hormone was achieved by utilizing antibody-labeled superparamagnetic particles to facilitate capture on a biospecific surface [45].

The work herein extends the concept of size-based AFM immunoassays through a preliminary investigation of pathogen detection that relies solely on the size of the selectively captured analyte on a chip-scale platform. We then describe an exploration of the use of different-sized particles of immunogold conjugates as labels in a sandwich-type immunosorbent assay. We conclude with an assessment of the performance of this method, along with discussions of strategies for improvements in throughput and sensitivity.

4.2. EXPERIMENTAL SECTION

4.2.1. Reagents

Potassium phosphate, potassium chloride, boric acid, sodium chloride, and tetrahydrofuran were purchased from Fisher and used as received. Tween 80, bovine albumin, and dithiobis(succinimidyl propionate) (DSP) were obtained from Sigma. Polyclonal antibodies of rabbit IgG, rat IgG, goat anti-rabbit IgG, and goat anti-rat IgG were used as received (Pierce). Gold-conjugated goat anti-rabbit (10 nm) and goat anti-rat (30 nm) were purchased from Ted Pella. Octadecanethiol (ODT) was acquired from Aldrich and recrystallized from ethanol (Quantum, punctilious grade). Poly(dimethylsiloxane) (PDMS) was obtained from Dow Corning. Contrad 70, used for cleaning glass was purchased from Decon Labs, and two-part epoxy, employed for making template stripped gold, was obtained from Epoxy Technology. Borate buffer packs were purchased from Pierce, and all buffers were filtered with a 0.22-μm syringe filter (Costar) before use. Dithiobis(succinimidyl undecanoate) (DSU) was synthesized from a combination of literature procedures [67,68].

FCV (titer 5×10^8 viruses/mL via the Reed and Muench $TCID_{50}$ method [69]) was obtained in cell lysate from the National Animal Disease Center (NADC) and used as received. Anti-FCV mAbs were also supplied by NADC after purification with a protein G column to 99.9% (Bio-Rad) and stored in 10 mM phosphate buffered saline (PBS).

4.2.2. Gold Substrate Fabrication

Template-stripped gold (TSG) surfaces were prepared using the method of Wagner and coworkers [70,71]. First, a 300-nm layer of gold (99.99%) was vacuum deposited onto freshly cleaved mica (Asheville-Schoonmaker) or a *p*-type silicon wafer [University Wafer 4″ (111)] at a rate of 0.1 nm/s in a cyro-pumped Edwards 306A evaporator while maintaining a chamber pressure below 5×10^{-6} Torr. The substrates prepared on mica were used for the multiplexed sandwich assays, while those prepared on silicon were used for the virus assays. The mica-derived substrates were heated prior to gold deposition at 150°C for ~18 h to remove adsorbed volatile contaminants.

The next step sandwiched ~10 μL of Epo-tek 377 epoxy between the gold substrate and a 1-cm^2 section of glass previously cleaned in dilute surfactant solution (Contrad 70), and then sonicated in deionized water and ethanol for 30 min each. The sandwiched samples were then annealed at 150°C for 90 min. The final step consisted of immersing the annealed mica samples in tetrahydrofuran for 5–15 min, which led to separation of the gold surface from the mica backing. Silicon wafer samples did not require immersion in tetrahydrofuran and the glass chips were simply detached from the wafer with tweezers to expose the underlying gold-coated glass substrate. After separation from the backing material, the TSG surfaces were then rinsed with ethanol and dried under a stream of argon. This process yields an extremely flat gold surface (see below).

4.2.3. Preparation of Capture Antibody Surfaces

Capture antibody substrates were prepared in several steps. First, a PDMS stamp was constructed with a ~0.5-cm-diameter hole in the center and soaked in a 1 mM ODT solution for 30 s. Next, the TSG substrates were exposed to the ODT-soaked stamp for ~30 s [72–74]. The ODT-inked TSG surface was rinsed with ethanol, and then immersed in 100 μM ethanolic solution of DSU or 2-mM ethanolic solution of DSP. Immersion times were typically ~12 h for both. These steps resulted in the chemisorption of DSU or DSP as its corresponding thiolate to the central, unmodified region of the TSG chip. This procedure provides a hydrophobic barrier around a succinimidylester-terminated monolayer that can be used to covalently couple antibodies in a small, localized area [75–78]. Previous studies have demonstrated the ability of DSU- and DSP-based monolayers to covalently bind primary amine containing molecules (e.g., the lysine residues of an IgG antibody) to surfaces [52,67,68]. The DSU-derived surfaces were used in the multiplexed sandwich assays, whereas the DSP-derived surfaces were employed in the virus assays.

For the virus assays, the DSP-derived surfaces were modified with a coating of anti-FVC mAbs. Anti-FCV mAbs were covalently immobilized to the DSP-modified substrates by pipetting 20 μL of the protein solution [100 μg/mL in 50 mM borate buffer (pH 8.5)] onto the substrate and reacting for ~7 h at room temperature; these substrates were rinsed 3 times with 10 mM PBS (pH 7.4).

For the multiplexed sandwich assays, the two-component capture substrates were prepared by the codeposition of 50 μL of a 1:1 mixture of polyclonal goat anti-rabbit IgG and polyclonal goat anti-rat IgG in borate buffer solution (100 mM H_3BO_3, 100 mM KCl, 1% Tween 80) at pH 9.0 for ~6 h. The total IgG concentration was 0.1 mg/mL, and corresponded to ~100 times the saturation coverage of the substrate. After immobilization of the capture antibody layer, the samples were rinsed with deionized water and dried in a stream of argon. To block any unreacted DSU, a solution of bovine albumin (10 mg/mL) in borate buffer at pH 9.0 was localized on the surface for 1 h. The blocked antibody-modified surfaces were then rinsed with deionized water and dried under argon before immediate exposure to analyte solutions. A blocking step was not required in the virus assays.

4.2.4. Immunoassay Protocols

4.2.4.1. Virus Assays. After the substrates were rinsed with 10 mM PBS, 20 μL of the virus solution was pipetted onto the substrates and incubated at room temperature for 16 h. The substrates were then rinsed 3 times by 2.5 mL volume displacement with 10 mM PBS (pH 7.4), and once with a 3-s stream of deionized water to remove the residue from the high salt solutions prior to AFM imaging. Finally, the substrates were dried with a stream of nitrogen. Dose–response curves were constructed by diluting FCV to varying concentrations in cell lysate and collecting AFM images for enumeration.

4.2.4.2. Nanoparticle-Based Assays. The two-component capture substrates were prepared by exposure to 50 μL PBS (pH 7.6; 50-mM potassium phosphate, 150 mM NaCl) that contained either polyclonal rabbit IgG, rat

IgG, or mixtures of both for 60 min at room temperature; initial concentrations of IgG were determined by the absorbance at 280 nm. Samples were then rinsed with PBS and exposed for 65 min to 50 μL of a 1:1 mixture of 10-nm gold particles labeled with polyclonal goat anti-rabbit IgG and 30-nm gold particles labeled with polyclonal goat anti-rat IgG. A small amount of Tween 80 (1%) was added to the particle suspension, which was buffered at pH 8.2 (20 mM Tris, 20 mM sodium azide, 225 mM sodium chloride, 1% BSA, 20% glycerol). The final concentrations of the 10- and 30-nm gold conjugates were 8.5×10^{12} and 4×10^{11} particles/mL, respectively. These samples were rinsed with a high salt phosphate buffer (2.5 M NaCl) to minimize nonspecific ionic attractions between the negatively charged gold particles and substrate surface. The samples were then rinsed with deionized water and dried under argon.

4.2.5. AFM Imaging

A MultiMode NanoScope IIIa AFM (Digital Instruments), equipped with either a 12- or 150-μm tube scanner and operated under ambient conditions in tapping mode, was utilized for all measurements. Images were collected at scan rates of 1–2 Hz, while maintaining a constant setpoint amplitude for cantilever oscillation through a feedback loop. The amplitude of the setpoint oscillation was maximized relative to the amplitude of free oscillation in order to minimize the force applied to the samples. The probe tips were n(+)-silicon TESP cantilever/tips (nanosensors) with a length of 118 nm, a width of 27–29 nm, and thickness of 3.6–4.5 nm; force constants between 38.5 and 72.4 N/m; and resonance frequencies, between 298 and 421 kHz.

Three-five discrete regions were imaged on all samples. The number of particles per square micrometer was determined either through the use of particle-counting software (Scion Image, Scion Corporation) or by direct enumeration with a pen-style colony counter (Sigma). We did not observe any appreciable self-aggregation of gold conjugates by transmission electron microscopy (data not shown); hence, each particle bound was counted regardless of whether it was fully resolved from neighboring particles. Typical scan areas were 2 × 2 μm, 5 × 5 μm, or 10 × 10 μm, depending on particle density.

4.2.6. Infrared Spectroscopy

Infrared reflection spectra (IRS) were acquired using a Nicolet 750 FTIR spectrometer purged with liquid N_2 boiloff and a liquid N_2-cooled HgCdTe detector. IRS spectra were obtained using *p*-polarized light incident at 82° with respect to the surface normal. Octadecanethiolate-d_{37}-coated gold was employed as the reference. Sample and reference spectra are the average of 512 scans at 2 cm^{-1} resolution (one level of zero filling) with Happ–Genzel apodization.

4.3. RESULTS AND DISCUSSION

4.3.1. Supporting Substrate

To successfully detect the topographic changes associated with the specific binding of nanometer-sized analytes and labels, it is critical that the roughness of the underlying substrate be at least 3 times lower than that of the size of the imaged objects. To meet this requirement, we have taken advantage of the relatively low roughness of template stripped gold (TSG) surfaces over micrometer-length lateral dimensions [71]. Figure 4.1 presents topographic AFM images ($2 \times 2\,\mu m$ and $500 \times 500\,nm$) of an unmodified TSG surface. The surface has an rms roughness of 0.2 nm with variations in topography of only ~4 nm over a $10 \times 10\,\mu m$ area. This level of roughness is well below that of the smallest label used in any of the assays (i.e., the 10-nm gold particles employed in the multiplexed immunoassay), indicating that TSG surfaces will more than effectively function for our purposes as capture antibody substrates. Furthermore, the low surface roughness allows detection of labels as small as 0.6 nm, resulting in a plethora of potential label sizes.

4.3.2. Capture Substrate

Several different capture substrates were prepared by the immobilization of monoclonal and polyclonal antibodies on TSG by using either DSU or DSP as a coupling agent (Fig. 4.2). To characterize the effectiveness of these procedures and to assess the effect of immobilization on the roughness of the resulting surface, two sets of experiments were conducted. The first used AFM to monitor the density of bound antibody as a function of derivatization time (Fig. 4.3), and the second followed the modification process by infrared reflection spectroscopy (IRS) (Fig. 4.4).

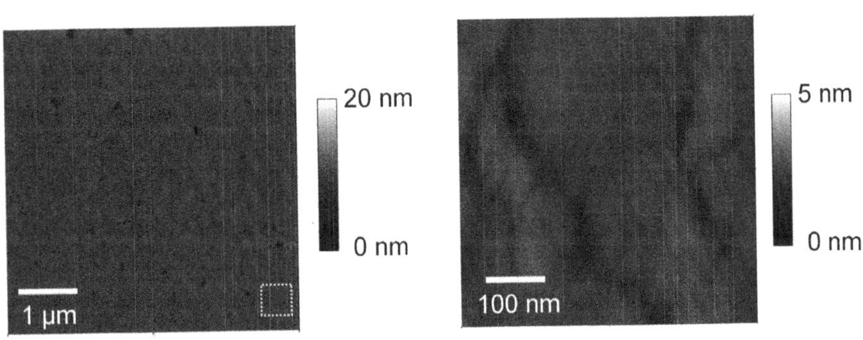

Figure 4.1. AFM topography images ($5 \times 5\,\mu m$ and $500 \times 500\,nm$) of an uncoated TSG surface. The $500 \times 500\,nm$ image is a zoomed in image of the dotted box in the $5 \times 5\,\mu m$ micrograph. The rms roughness factor of the substrate is 0.2 nm.

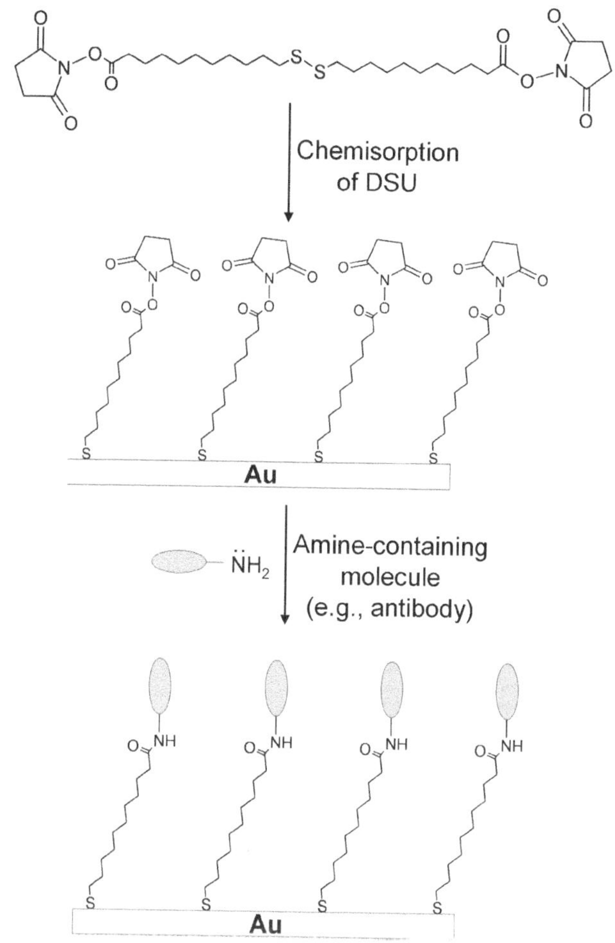

Figure 4.2. Idealized scheme for the preparation of capture substrates. TSG is modified with a DSU-derived monomolecular layer, followed by the covalent coupling of an amine-containing molecule.

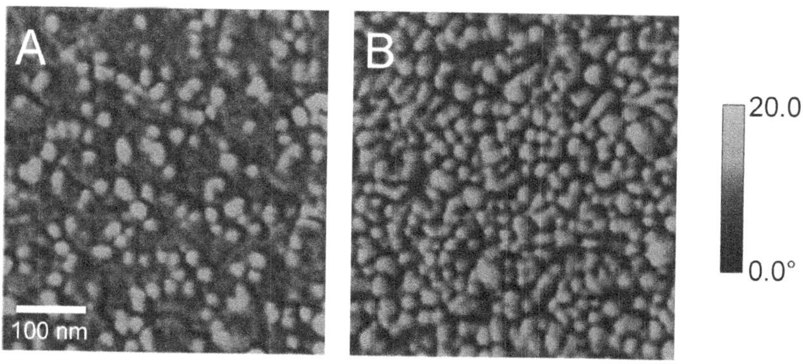

Figure 4.3. AFM phase images (500 × 500 nm) of DSU/TSG surfaces after (a) 1 h and (b) 4 h exposure to capture antibody solutions.

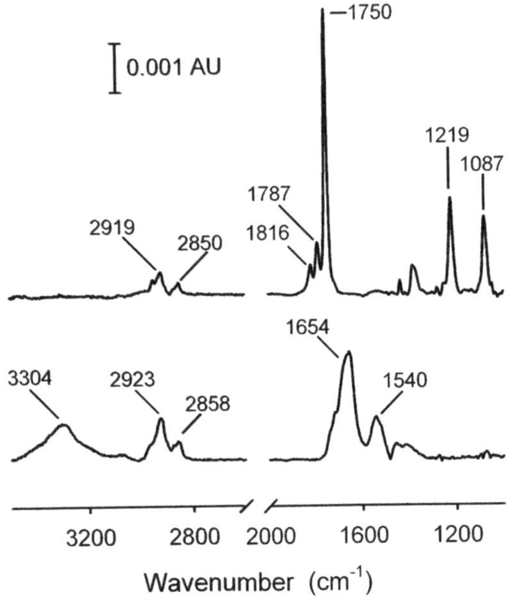

Figure 4.4. IRS verification of the formation of a DSU-derived monolayer and the presence of antibody on the substrate.

Figure 4.3 shows phase images of DSU-modified surfaces after exposure to a mixture of capture antibodies (IgG proteins) for 1 h (a) and 4 h (b). Phase imaging was used in this characterization in order to partially compensate for local variations in surface topography. Phase imaging also provides an increase in contrast because of differences between the viscoelastic properties of the IgG protein and surrounding DSU-based monolayer [79–81]. This capability is evident in Figure 4.3. Immobilized IgG antibodies are visible as small, isolated features that can be ascribed to immobilized IgG showing lateral cross section of ~25 nm. There is also evidence for a few small aggregates. The increase in lateral dimensions from that for IgG by X-ray crystallography (14 × 8.5 × 5 nm) [82] is due to tip convolution effects, and agrees with earlier reports on the imaging of antibodies with AFM [44,83–85].

Figure 4.3a shows that a low coverage layer is formed after 1 h, with only ~50% of the surface coated. As the exposure time is increased to 4 h, the surface approaches a spatially limited packing density that is evident in Figure 4.3b. This image reveals a much more densely packed antibody-modified surface. Hence, the DSU- or DSP-modified TSG surfaces were exposed to capture antibody solutions for a minimum of 5.5 h in all subsequent experiments to ensure that a fully coated capture substrate was formed.

The modification of the substrate was also monitored by IRS. Figure 4.4 shows several spectra that illustrate the reaction between the DSU terminus and amine-containing molecules. For the DSU-based monolayers, several bands characteristic of the succinimidyl functionality are apparent: the carbonyl stretch of the ester (~1820 cm^{-1}); in-phase (~1788 cm^{-1}) and out-of-phase (~1750 cm^{-1}) carbonyl stretches of the succinimidyl group; the symmetric (~1370 cm^{-1}) and asymmetric (~1210 cm^{-1}) C—N—C stretch; and the N—C—O stretch of the succinimidyl group (~1070 cm^{-1}). These peaks confirm the presence of a DSU-formed monolayer; comparable results were found for the DSP-derived monolayer [86].

Investigations into the binding efficiency of the DSU-derived monolayer toward IgG proteins using IRS reveal the presence of infrared absorption bands diagnostic of a secondary amide at 1653 cm^{-1} (amide I) and 1555 cm^{-1} (amide II). Most importantly though, is the absence of the bands of the DSU succinimidyl endgroup.

Figure 4.4 reveals that the bands corresponding to the succinimidyl endgroup are still present at 20–25% of their original intensities indicating that not all of the succinimidyl terminal end groups have formed covalent linkages to the protein [87]. Any amide stretches that are formed by the nucleophilic attack of the amine group of the protein on the carbonyl carbon of the adlayer terminus, unfortunately, are obscured by the presence of the same vibrational modes in IgG. Incomplete conversion of the terminus is not unexpected, due to the much larger cross section of the antibody (~120 nm^2) versus that of the endgroup of the adlayer(~2 nm^2) [67]. To minimize the presence of any unreacted terminal groups, the samples were exposed to a 1 mg/mL solution of bovine serum albumin (BSA) for 60 min after coupling of the capture antibody. As is evident from control experiments devoid of analyte (see next section), the nonspecific binding of the gold conjugates is relatively low, indicating efficient blocking of unreacted DSU.

4.3.3. Virus Assays

4.3.3.1. Scope. This section demonstrates the potential of AFM as a readout tool for the detection of pathogens (Scheme 4.1). These experiments also demonstrate a potential combinatorial approach to screen several different monoclonal antibodies for effectiveness, and detailed calibration curves were established to determine limits of detection. Results of tests for binding specificity, incubation time, and the effect of blocking agents will be reported elsewhere.

As a first demonstration of a size-based assay, FCV was chosen as a model because it is nanometrically sized (36 nm diameter), spherically shaped and mechanically noncompliant [88]. These characteristics allow direct image identification without use of labels. The well-defined structure of the FCV also enhances the ability to distinguish its presence from other nonspecifically bound material. We add that FCV serves as a surrogate for the Norwalk-like virus (NLV) [89], the human strain of calicivirus, which is responsible for many food/water-related illnesses and has been identified as a potential bioterrorism agent [90,91]. Thus, the use of FCV represents a well-defined starting point for assessments of the potential of AFM as a diagnostic tool.

4.3.3.2. Dose–Response Curves. AFM images were collected at varying concentrations of FCV to evaluate the effectiveness of the method. A 25 μm^2 scan size was chosen to best showcase the FCV. Tests indicated that this scan size serves as a representative sampling area while maintaining sufficient image resolution to accurately distinguish virus particles from other nonspecifically bound material. As evident in Figure 4.5, FCV appears as spherical objects with a height of ~22 nm. The height is somewhat smaller than expected due to dehydration from air-based imaging. In addition, the AFM images have been flattened with software to facilitate analysis, which further diminishes the apparent height of the objects. Most importantly, all FCV appear at the same height (22 nm), which enables facile quantification. Irregularly shaped objects can also be seen in the images and are attributed to cell debris and other components in the spent cell culture media. Figure 4.5 also shows that

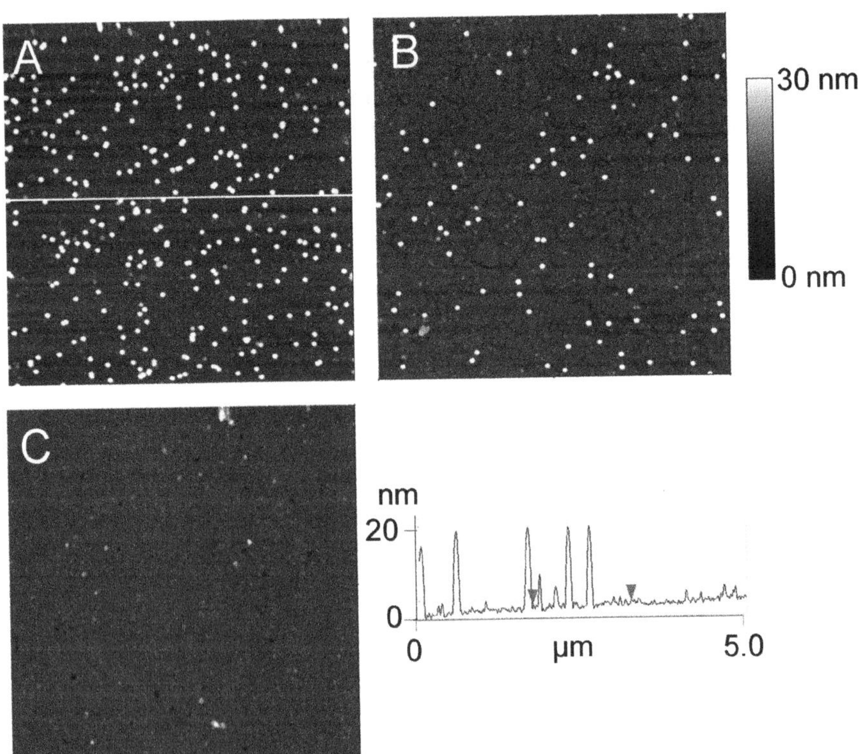

Figure 4.5. AFM micrographs (5 × 5 µm) of FCV bound to capture substrates at three different FCV concentrations: (a) 3×10^8, (b) 5×10^7, and (c) 0 viruses/mL. FCV was dispersed in cell lysate.

the number of captured FCV is concentration-dependent, implying that the method has potential utility as a quantitative analytical tool. However, the images also depict the FCV having varying lateral dimensions. This observation is attributed to tip convolution effects. Many of the images shown were collected with AFM tips that have differing characteristics (e.g., shape and sharpness) that change the apparent lateral dimensions of imaged objects [92–94]. One advantage of AFM is, however, its ability to distinguish objects on the basis of topographic changes, regardless of tip convolution effects. As the cross sectional plot in Figure 4.5 documents, FCV has the same average height (~22 nm). A more detailed analysis of these data will be presented shortly from a quantitative perspective.

To optimize the immunoassay, three different mAbs for FCV (NADC strains 2D10-1C4, 4G1-2F4, and 3B12-1C8), were screened for effectiveness by exposure of FCV to capture substrates of each mAb. AFM images from these experiments, on exposure to a virus concentration of 3×10^8 viruses/mL, are shown in Figure 4.6. It is readily apparent from these images that antibody 4G1-2F4 has significantly less binding than the other two mAbs.

To further this assessment, each of the capture substrates were tested five or more times at several different virus concentrations, with five images collected at different locations on each sample. The number of FCV bound in a $25\,\mu m^2$ image was averaged and standard deviations were calculated for each FCV concentration. The resulting dose–response curves are shown in Figure 4.7. The slopes of the dose–response curves are related to the binding affinities of the mAbs, and therefore can be used to identify the antibody with the highest affinity for FCV. Test solutions covered the range of 3×10^8 to 5×10^6 viruses/mL and were diluted in cell lysate to mimic a biological matrix. Figure 4.7 shows that the most sensitive and effective mAb is 2D10-1C4, while the least sensitive mAb is 4G1–2F4. For example, when 3×10^8 viruses/mL were exposed to the different antibodies, 2D10-1C4 bound 311 viruses in $2\,\mu m^2$, while 4G1-2F4 bound only 51 viruses. Anti-FCV mAb 2D10-1C4 therefore shows the most effective binding. AFM imaging therefore provides a fast and simple method for determining the quality of mAbs for FCV and can be applied to many other screening applications.

From the dose–response curves in Figure 4.7, limits of detection for each mAb were determined. The limit of detection is defined as the concentration of virus that can be determined above the average blank signal plus three times the standard deviation of the blank samples. The blank signal is attributed to objects in solution that have a size and shape comparable to FCV. These data translate to a limit of detection of 3×10^6 for 2D10-1C4, 7×10^6 for 3B12-1C8, and 3×10^7 viruses/mL for 4G1-2F4. These results are consistent with expectations based on binding affinity (i.e., the slopes of the plots in Fig. 4.7). These detection limits are comparable to those of other viral detection schemes, such as fluorescence [95] and microcantilevers [96], which have detections limits between 10^5 and 10^8 viruses/mL. More importantly, these results demonstrate the potential of AFM imaging as a sensitive, size-based

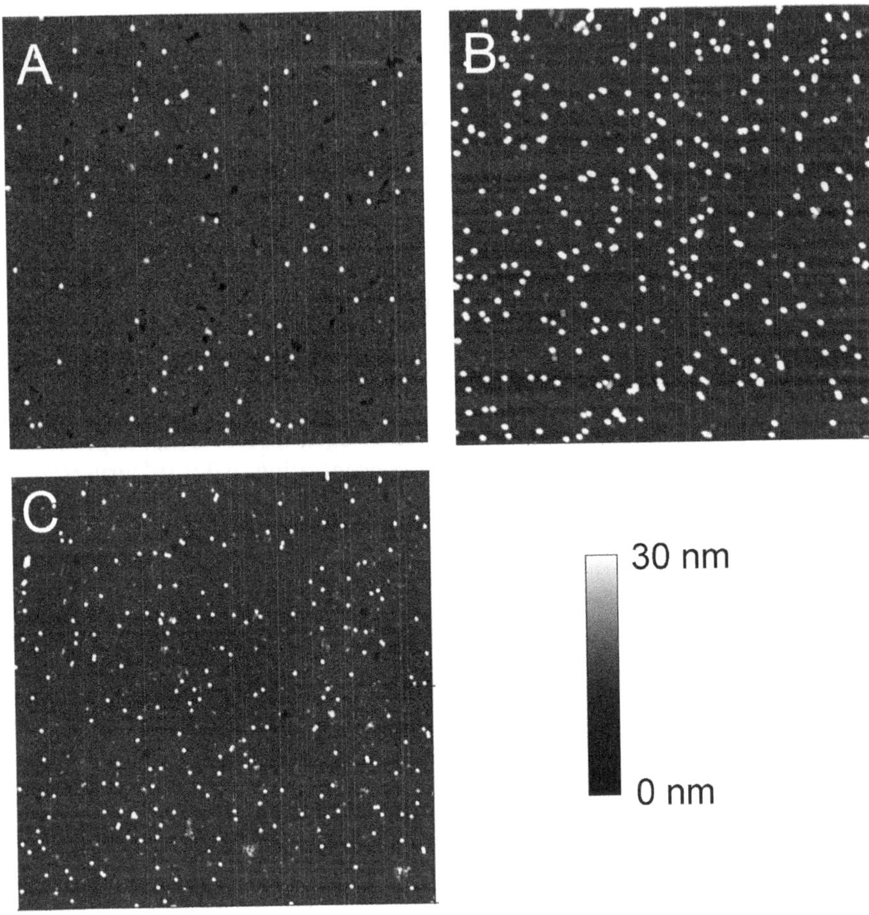

Figure 4.6. AFM micrographs ($5 \times 5\,\mu m$) of FCV captured by substrates modified with different anti-FCV mAbs. Each image was obtained after exposure of a 3×10^8 viruses/mL solution in cell lysate on a substrate modified with either mAb (a) 4G1-2F4, (b) 2D10-1C4, or (c) 3B12-1C8.

diagnostic tool for pathogen detection; an approach easily adapted to combinatorial assessment via addressing or differentiation based on pathogen size.

4.3.4. Size-Based Label Assays

4.3.4.1. Scope. This section extends the assay based on size to much smaller analytes, i.e.; IgG proteins by use of 10- or 30-nm gold particles conjugated to a labeling antibody in a sandwich type assay. This section also examines the possibility of using 10- and 30-nm gold particles as labels in a dual-analyte sandwich format. Scheme 4.2 depicts both aspects of the concept.

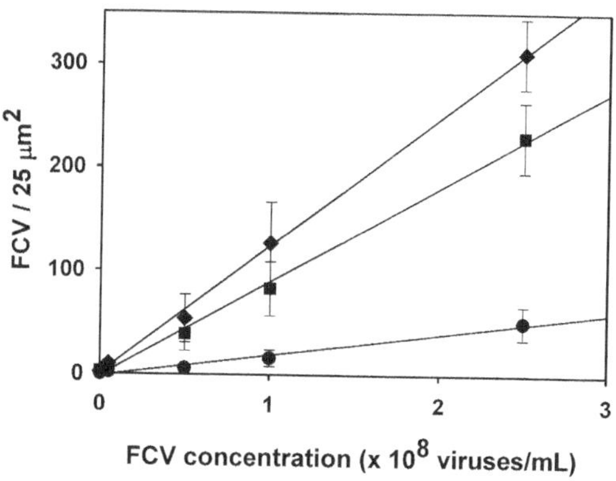

Figure 4.7. Dose–response curve responses for three different mAbs substrates exposed to FCV for ~16 h. The dose–response curves yield limits of detection of 3×10^6 viruses/mL (2D10-1C4, △), 7×10^6 viruses/mL (3B12-1C8, ■) and 3×10^7 viruses/mL (4G1-2F4, •).

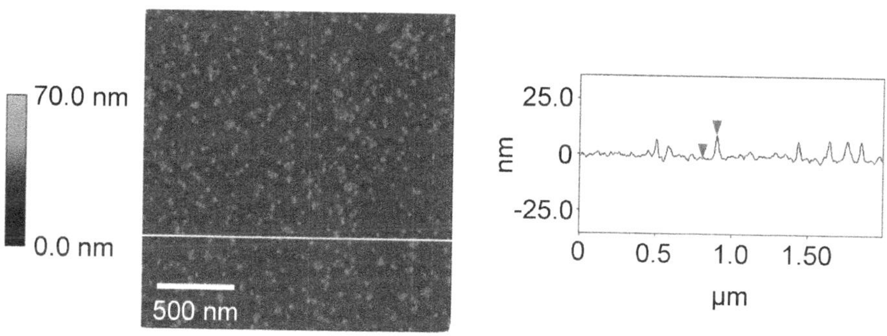

Figure 4.8. AFM topography image (2 × 2 μm) of a capture antibody-modified surface after exposure to 10 μg/mL rabbit IgG and the mixed gold conjugate suspension. A cross section through the center of a captured 10-nm goat anti-rabbit gold conjugate, depicted by the white line in the micrograph, is shown. The difference in height between the two demarcation arrows is 10.5 nm. The conjugated particle concentrations were 8.5×10^{12} particles/mL (10 nm) and 4×10^{11} particles/mL (30 nm).

Figure 4.8 shows a 2 × 2 μm topographic image for a TSG substrate modified with two different IgG antibodies (anti-rat and anti-rabbit) after exposure to a solution containing only rabbit IgG and then to a suspension of both sizes of labeled nanoparticles. Thus, the resulting AFM image should be composed only of 10- and not 30-nm particles, reflecting the specificity of the goat anti-rabbit IgG layer coated on the 10 nm particles to bind with captured rabbit IgG. Indeed, the image reveals the presence of a large number of similarly sized particles. A cross sectional analysis of the surface (Fig. 4.8) reveals heights of 9–12 nm and lateral diameters on the order of 20–25 nm for captured particles. As before, the larger than expected lateral diameter of the particles is a consequence of distortions due to the size and shape of the probe tip [92–94]. Interestingly, the presence of the antibody coated on the particle does not appear to substantially contribute to the measured topographic change. More importantly, the surface is devoid of 30-nm particles, which is indicative of little (if any) nonspecific adsorption of the labeled 30-nm particles.

The same type of capture substrate was exposed to a rat IgG solution and subsequently to the mixed particle suspension. A representative image (2 × 2 μm) is presented in Figure 4.9. Since only rat IgG was present in the sample, only the 30-nm particles conjugated with goat anti-rat IgG should be captured. This expectation is consistent with the cross sectional plot in Figure 4.9; that is, the surface has captured particles with heights of 26–31 nm. Again, very little nonspecific binding of 10-nm particles conjugated with goat anti-rabbit is observed. This set of results further verifies the effectiveness of the assay, highlighting the ability of AFM to function as an intriguing readout modality for a chip-based assay.

Based on these findings, a first-generation test of the possible multiplexing ability of this strategy was conducted. The image in Figure 4.10 represents a capture substrate with two different antibodies exposed first to a solution containing both antigens at the same concentration, and then to the mixed particle suspension. In this case, both 10- and 30-nm particles are observed, diagnostic of the presence of both rabbit IgG and rat IgG in the sample. The cross section in Figure 4.10 shows that the relative differences in size between the 10- and 30-nm particles are easily distinguished by AFM against the background topography of the capture substrate. It is also apparent that for the same approximate concentration in solution, the number of 10-nm particles captured by the substrate is much higher than that of the 30-nm particles. We

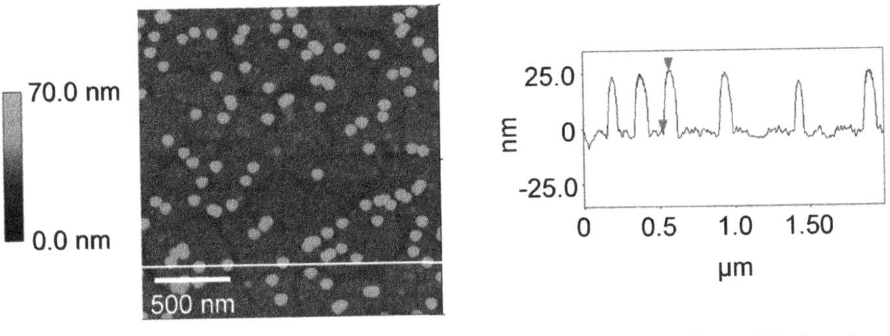

Figure 4.9. AFM topography image (2 × 2 μm) of a capture antibody-modified surface after exposure to 1 g/mL rat IgG and the mixed gold conjugate suspension. A cross section through the center of a bound 30-nm goat anti-rat gold conjugate depicted by the white line in the micrograph is shown. The difference in height between the two arrows is ~28 nm.

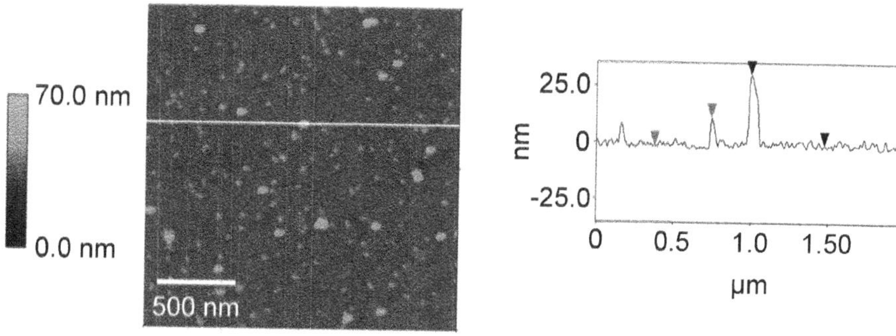

Figure 4.10. AFM topography image ($2 \times 2\,\mu m$) of a capture antibody-modified surface after exposure to a solution of 500 ng/mL rabbit IgG and 500 ng/mL rat IgG, and then to the mixed gold conjugate suspension. A cross section through the center of bound 10-nm and 30-nm conjugates depicted by the white line in the micrograph is also shown. The heights of the conjugates are ~11 and ~31 nm, respectively.

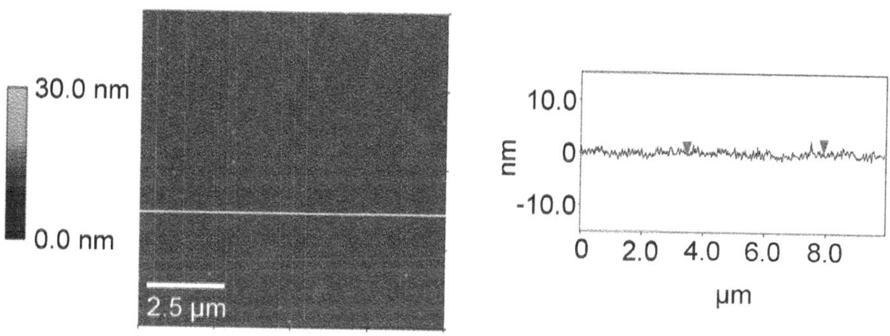

Figure 4.11. AFM topography image ($10 \times 10\,\mu m$) of a capture antibody-modified surface after exposure to a blank PBS buffer solution and the mixed gold conjugate suspension. A cross section demonstrating the low roughness of the capture antibody/DSU/TSG surface depicted by the white line in the micrograph is also shown.

believe this result is due to two different contributing factors, which will be discussed shortly.

A critical factor in determining the detection limit of a two-site "sandwich" solid-phase immunoassay is the nonspecific adsorption of detection antibody to the substrate surface in the absence of analyte. Ideally, no detection particles should be observed when there is no analyte in solution. The topographic image in Figure 4.11 represents for a $10 \times 10\,\mu m$ area of a capture substrate exposed to buffer solution devoid of rabbit IgG and rat IgG, followed by the particle suspension. In this image, only fourteen 10-nm particles and seven 30-nm particles are observed. The corresponding image cross section in Figure 4.11 again demonstrates the low level of roughness of the capture substrate, enabling the sensitive detection of any particles present on the surface. This low level of nonspecific binding is advantageous when attempting to detect the presence of low levels of analyte in complex biological media. However, experiments along these lines remain to be explored.

4.3.4.2. Quantitative Dose–Response Behavior: Size-Based AFM Assay. The previous section demonstrated the ability to concurrently detect two analytes within micrometer-sized regions of a surface using a size-based AFM assay. The next issue to address is what type of quantitative information can be obtained for each analyte in such a system, and what effects are inherent because of the different sized particles used for analyte identification. We first investigated the differences in dose–response behavior for each analyte individually in order to establish any differences in the sensitivity when using different-sized particles.

A dose–response curve for rabbit IgG, which binds 10-nm goat anti-rabbit particles, is shown in Figure 4.12a. The dose–response curve initially rises with increasing rabbit IgG concentration, leveling off at ~1000 ng/mL. This plateau is likely due to saturation of available antigenic sites on the surface, as only a ~10% increase in bound 10-nm particles is observed for a 10-fold increase in concentration to 10,000 ng/mL. Furthermore, in data not shown, the number of bound particles is indistinguishable at concentrations below 25 ng/mL, although the average response remains at a level at least three times above the measured background down to a limit of detection of 5 ng/mL.

A comparison of the relative labeling efficiencies of the two different-sized particles can be established by examination of the dose–response data in Figure 4.12b for rat IgG, which binds 30-nm goat anti-rat particles. The plots indicate an extension of the dynamic range to ~5000 ng/mL for rat IgG; however, the sensitivity is reduced to ~10% of that for rabbit IgG response. Examination of the data in Figure 4.12b yields a detection limit of ~50 ng/mL for rat IgG, assigning a signal-to-background ratio of 3:1 as the lower limit of detection.

4.3.4.3. Dual-Analyte Response for Sized-Based AFM Assay. To investigate the effect of the competition of different-sized particles when two different analytes are present, three mixed capture antibody surfaces were prepared and exposed to differing mixtures of rabbit IgG and rat IgG (Table 4.1). For example, the response for 500 ng/mL rat IgG in the absence of rabbit IgG is 6.71 ± 1.62 particles/μm^2, but is reduced by ~30% (4.84 ± 0.92 particles/μm^2) in the presence of 500 ng/mL rabbit IgG. Similarly, the response for 500 ng/mL rabbit IgG in the absence of rat IgG is 79.92 ± 6.25 particles/μm^2, but is decreased by ~45% to 46.6 ± 4.87 particles/μm^2 in the presence of 500 ng/mL rat IgG. These data suggest that there is a steric hindrance in the presence of a competing analyte that leads to lower signals.

The difference in sensitivity for the rat versus rabbit IgG has several possible origins. One possibility is a differences in the binding affinities of rabbit IgG versus rat IgG. However, an estimate of absolute binding affinity is

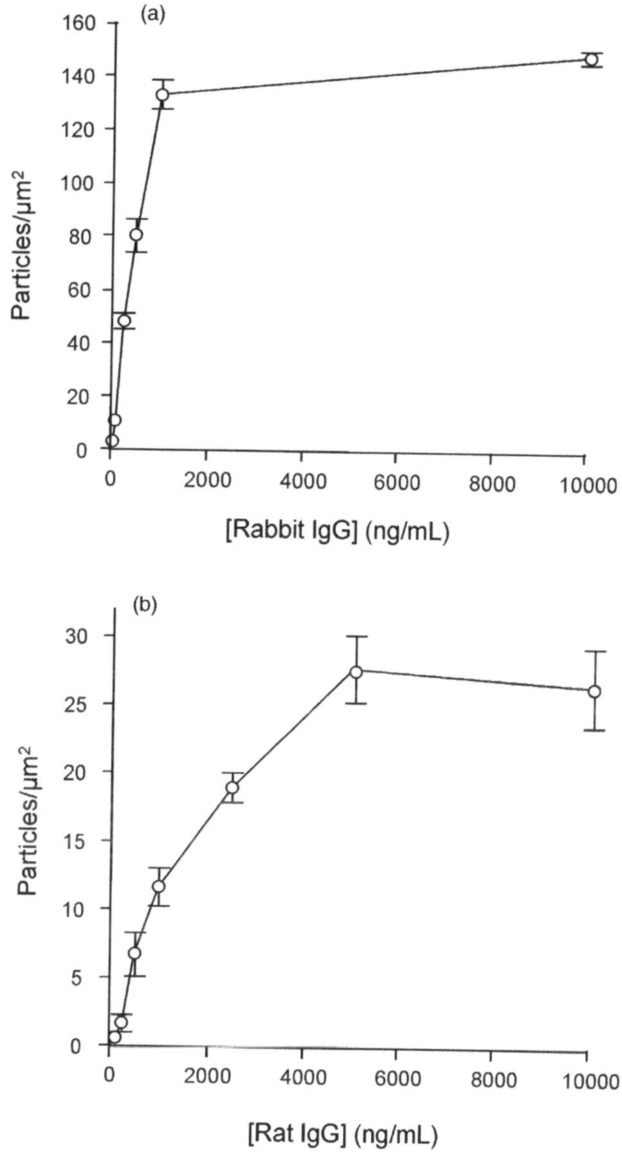

Figure 4.12. Dose–response curves for rabbit (a) and rat (b) IgG.

difficult to ascertain because of the polyclonal nature of these antibodies. To investigate the possibility of differences in binding affinity, surfaces were exposed to a particle suspension in which the conjugate sizes were reversed, i.e., 10-nm particles were conjugated to goat anti-rat IgG and 30-nm particles were conjugated to goat anti-rabbit IgG. A similar decrease in the number of captured 30-nm particles for rabbit IgG relative to 10-nm particles for rat IgG at the same analyte concentration was observed. This result verifies that the

TABLE 4.1. Dual analyte responses for rabbit- (10-nm particles) and rat IgG (30-nm particles)

Sample	30-nm (particles/μm²)	10 nm (particles/μm²)
500 ng/mL rat IgG + 50 ng/mL rabbit IgG	4.94 ± 1.82	1.82 ± 0.15
500 ng/mL rat IgG + 250 ng/mL rabbit IgG	4.16 ± 1.31	23.05 ± 3.97
500 ng/mL rat IgG + 500 ng/mL rabbit IgG	4.84 ± 0.92	46.60 ± 4.87
250 ng/mL rat IgG + 500 ng/mL rabbit IgG	1.49 ± 0.33	40.00 ± 4.85
50 ng/mL rat IgG + 500 ng/mL rabbit IgG	0.09 ± 0.02	45.94 ± 4.40

differences in sensitivity are due to differences in particles and not binding affinity.

Another possible origin of the decreased response for rat versus rabbit IgG is the larger cross section of a 30- versus 10-nm particle. Neglecting the antibody coating, a 30-nm particle will occupy a surface area of ~2800 nm^2 on capture, whereas the footprint of a 10-nm particle is only ~310 nm^2. As a consequence, a 30-nm particle will sterically hinder access to more neighboring antigenic sites on the capture surface than a 10-nm particle, which would lower the sensitivity for detection by the 30-nm particles. Previous AFM studies [48,52,58,97] have observed thicknesses on the order of 3–5 nm for a monolayer of bound IgG antibody. When considering the dimensions of IgG from X-ray crystallography (14 × 8.5 × 5 nm), it is likely that the IgG antibody is lying flat on the surface, therefore occupying a cross section of ~120 nm^2. This situation would result in a 10-nm particle blocking up to ~three antigenic sites while a 30-nm particle could limit access up to ~25 antigenic sites, accounting for, at least in part, the observed differences in sensitivity.

In addition to steric effects, the diffusional characteristics of the two different sized labels should be considered. Earlier work investigated different factors affecting the staining efficiency of colloidal gold particles for use in the electron microscopic determination of surface concentrations of a target protein [98]. By assuming that the binding of a gold particle to a surface bound receptor is a diffusion-limited process, Equation (4.1) estimates the number of particles impinging on a unit surface area (q, particles/cm^2) at time, t via Einstein's law of Brownian motion:

$$q = 0.163n(kTt/Vr)^{1/2} \tag{1}$$

In this equation, n is the concentration of gold particles in bulk solution (particles/mL), k is Boltzmann's constant (1.38×10^{-16} g·cm^2/s^2·K), V is the viscosity of the medium (g/cm·sec), T is temperature (K), and r is the radius of the gold particle (cm). This equation assumes a sticking coefficient of unity, i.e., every labeling particle that encounters the surface binds. This assumption is not likely valid in the presence of low analyte concentrations where only a small amount of analyte is bound, decreasing the number of binding sites on the substrate. Therefore, the likelihood of a label encountering a binding site is significantly reduced. Nevertheless, it provides a basis to qualitatively asses the labeling efficiency of the different sized particles.

Equation (4.1) predicts that the largest influence on labeling efficiency will arise from (1) the concentration of gold particles in bulk solution and (2) the radius of the particle. If we assume all other conditions to be invariant for both particle sizes, q is proportional to n/r. Since the concentrations of gold particles used for labeling bound antigen were 8.5×10^{12} 10 nm particles/mL and 4×10^{11} 30-nm particles/m, the 10-nm particle will strike the capture substrate ~65 times more than 30-nm particles per unit time, further accounting for the observed differences in sensitivity.

Based on the preceding considerations, we believe that the observed dual-analyte response is the result of interplay of both constraints. While the 10-nm particles impinge upon the surface at a greater flux than the 30-nm particles, a captured 30-nm particle blocks access to a greater proportion of antigenic sites.

4.3.5. Strategies to Reduce Total Analysis Time and to Lower Limits of Detection

To this point, we have focused primarily on an assessment of the utility of AFM to function as a readout tool for chip-scale diagnostic platforms. However, there are two principal issues that need to be addressed in order to fully realize the potential of AFM as a valuable, high-throughput screening technique. The first deals with the time required for an analyte to equilibrate with a capture substrate. For simplicity, the work herein used 12-h incubations, which from a time dependence study is only a few hours longer than that required for equilibration of large objects like viruses and the nanoparticle labels. Thus, approaches to reduce the time for incubation are a top priority.

There are proven pathways to address this need, all of which are based primarily on mechanisms to enhance the rate of impingement of an analyte on the capture surface [99,100]. For example, operation at physiological temperatures increases the rate of diffusional mass transfer by raising translation energy and by lowering solution viscosity [99]. Preliminary tests with incubations at physiological temperature (e.g., 39°C for FVC) indicate that the time for equilibration is reduced to ~1 h. Moreover, a proof-of-concept experiment that delivered FCV to the capture substrate in a format that mimics the convective mass transfer process operative in rotating disk electrochemistry reached equilibrium in just over 10 min. It therefore appears a solution to lowering the time required for incubation, which has been a major limiting metric of performance is within reach.

The second critical issue rests with the limit of detection and readout time. As we have demonstrated, AFM can readily detect analytes at concentrations that are competitive with many other methods. However, further reductions in detection are required in order to be positioned to meet the eventual demands in areas like homeland security. It is not a question of image resolution. Several reports [34,35,101], including the work herein, have clearly documented the ability of AFM to image individual nanometer-sized objects one at a time. For AFM, the challenge arises from the low probability of finding one binding across a large sample area, which is further complicated by the low probability of an isolated target in a highly dilute solution striking the capture surface. In the case of the former, methods are needed to increase the ability to image a larger surface area more rapidly and to handle the huge increase in the resulting data files; although not yet solved, several research groups are pursing this objective [102,103]. With respect to the latter, mechanisms to rapidly concentrate a sample are required, with concepts based on electrophoretic [104,105] and magnetic focusing [106] beginning to emerge as potential solutions. Collectively, breakthroughs in these areas promise increases in imaging rates of ~30 and concentration factors of several thousand, which will result in a directly proportionate lowering of the limit of detection.

4.4. CONCLUSIONS

We have demonstrated the detection of two analytes simultaneously within micrometer-sized regions of a surface using size as a basis for identification. Gold nanoparticles of differing sizes, specific for different analytes present in sample solution, are imaged with AFM to determine the presence and amount of each analyte. The implementation of physical size labels instead of chemical labels potentially avoids problems associated with other dual detection formats, such as spectral overlaps and quenching, or pH-dependent signal responses. While the current format of the size-based assay appears to involve competition between labels, we believe that further improvements will reduce particle label competition and increase detection limits.

There are several modifications that could be implemented to improve the observed decreases in sensitivity. For example, the concentrations of 10- and 30-nm particles could be manipulated in solution to ensure that their fluxes are similar, removing the bias due to diffusional properties. In addition, steric limitations could be alleviated by creating a surface whose capture antibody sites are spatially separated such that the binding of a 30-nm particle would not greatly mask adjacent binding sites. The latter could be accomplished through the construction of a mixed DSU/glycol-terminated monolayer, where the glycol-terminated moieties resist nonspecific protein adsorption [107]. Indeed, a recent study demonstrated the manipulation of antigen density on a surface using such a mixed monolayer approach [108]. While such a mixed monolayer surface may decrease the observed signals due to a lower density of antigenic sites, this may be offset by the incorporation of higher-affinity monoclonal antibodies. We are currently investigating the effect of such changes to the observed response behavior.

In addition to the use of size as a label, it may also be possible to use particles with a similar cross section but different shapes, which could be differentiated by AFM. For example, the synthesis and characterization of gold nanorods has been demonstrated, with widths of ~10 nm and aspect ratios ranging from 1 to 7 [109,110]. Thus, gold nanorods of differing aspect ratios could be modified readily with antibodies and combined with spheres to allow for miniaturized shape-based AFM immunoassays. The combination of such shape- or size-based detection strategies with methods currently being developed for micrometer-sized patterning of multiple addresses may allow for an increased throughput through parallel processing of many analytes with minimal amounts of sample. Efforts at expanding the amount of analytes screened are currently underway.

ACKNOWLEDGMENTS

J. R. K. gratefully acknowledges the support of a Phillips Petroleum Corporation Graduate Research Fellowship. The support of a Dow Chemical Fellowship to K. M. K. is also gratefully acknowledged. We also recognize J. C. O'Brien and J. Ni for helpful discussions during preparation of this manuscript. This work was supported in part by the Institute of Combinatorial Discovery, Iowa State University, the Office of Basic Energy Science, Chemical Sciences Division of the U.S. Department of Energy, and by a grant from USDA-NADC. The Ames Laboratory is operated for the U.S. Department of Energy by Iowa State University under Contract W-7405-eng-82.

REFERENCES

1. Szostak, J., *Chem. Rev.* **97**:347–509 (1997).
2. Borman, S., *Chem. Eng. News* **76**:47–54 (1998).
3. Watkins, K., *Chem. Eng. News* **80**:30–34 (2001).
4. Hewes, J. D., et al. *Combinatorial Discovery of Catalysts: An ATP Position Paper Developed from Industry Input*, Gaithersburg, MD, 1998.
5. Cawse, J. N., *Acc. Chem. Res.* **34**:213–221 (2001).
6. Braeckmans, K., et al. *Modern Drug Discov.* **6**:28–32 (2003).
7. Eskelinen, M. and Haglund, U., *Scand. J. Gastroenterol.* **34**:833–844 (1999).
8. Chomel, B. B., *J. Vet. Med. Ed.* **30**:145–147 (2003).
9. Binning, G., Quate, C. F., and Gerber, C., *Phys. Rev. Lett.* **56**:930 (1986).
10. Gosling, J. P., *Clin. Chem.* **36**:1408 (1990).
11. Hadfield, S. G., Lane, A., and McIllmurray, M. B., *J. Immunol. Meth.* **97**:153 (1987).

12. Michael, K. L., Taylor, L. C., Schultz, S. L., and Walt, D. R., *Anal. Chem.* **70**:1242 (1998).
13. Gutcho, S., and Mansbach, L., *Clin. Chem.* **23**:1609 (1977).
14. Gow, S. M., Caldwell, G., Toft, A. D., and Beckett, G. J., *Clin. Chem.* **32**:2191 (1986).
15. Wians, F. H., Dev, J., Powell, M. M., and Heald, J. I., *Clin. Chem.* **32**:887 (1986).
16. Hayes, F. J., Halsall, H. B., and Heineman, W. R., *Anal. Chem.* **66**:1860–1965 (1994).
17. Bordes, A.-L., Limoges, B., Brossier, P., and Degrand, C., *Anal. Chem. Acta* **356**:195–203 (1997).
18. Buckwalter, J. M., Guo, X., and Meyerhoff, M. E., *Anal. Chem. Acta* **298**:11–18 (1994).
19. Blake, C., Al-Bassam, M. N., Gould., B. J., Marks, V., Bridges, J. W., and Riley, C., *Clin. Chem.* **28**:1469–1473 (1982).
20. Paek, S.-H. and Kim, J.-H., *Biotechnol. Bioeng.* **51**:591 (1996).
21. Macri, J. N., Spencer, K., and Anderson, R., *Ann. Clin. Biochem.* **29**:390–396 (1992).
22. Ohkuma, H., Abe, K., Kosaka, Y., and Maeda, M., *Anal. Chem. Acta* **395**:265–272 (1999).
23. Lewis, J. C. and Daunert, S., *Anal. Chem.* **71**:4321–4327 (1999).
24. Vuori, J., Rasi, S., Takala, T., and Vaananen, K., *Clin. Chem.* **37**:2087 (1991).
25. Salmain, M., Vessieres, A., Brossier, P., and Jaouen, G., *Anal. Biochem.* **208**:117–124 (1993).
26. Sayo, H. and Hosokawa, M., *Chem. Pharm. Bull.* **32**:1675–1678 (1984).
27. Chen, F.-T. A. and Evangelista, R. A., *Clin. Chem.* **40**:1819–1822 (1994).
28. Sayo, H. Hatsumura, H., and Hosokawa, M., *J. Chromatogra.* **426**:449–451 (1988).
29. Ni, J., Lipert, R. J., Dawson, G. B., and Porter, M. D., *Anal. Chem.* **71**:4903–4908 (1999).
30. Park, H.-Y., Lipert, R. J., and Porter, M. D., *Proc. SPIE* 464–477 (2004).
31. Grubisha, D. S., Lipert, R. J., Park, H.-Y., Driskell, J., and Porter, M. D., *Anal. Chem.* **75**:5936–5943 (2003).
32. Driskell, J. D., Kwarta, K. M., Lipert, R. J., Porter, M. D., Neill, J. D., and Ridpath, J. F., *Anal. Chem.* **54**:6147–6154 (2005).
33. Takano, H., Kenseth, J. R., Wong, S.-S., O'Brien, J. C., and Porter, M. D., *Chem. Rev.* **99**:2845–2890 (1999).
34. Allen, S., Chen, X., Davies, J., Davies, M. C., Dawkes, A. C., Edwards, J. C., Roberts, C. J., Sefton, J., Tendler, S. J. B., and Williams, P. M., *Biochemistry* **36**:7457–7463 (1997).
35. Reich, Z., Kapon, R., Nevo, R., Pilpel, Y., Zmora, S., and Scolnik, Y., *Biotechnol. Adv.* **19**:451–485 (2001).
36. McHugh, T. M., Wang, Y. J., Chong, H. O., Blackwood, L. L., and Stites, D. P., *J. Immunol. Meth.* **116**:213 (1989).

37. Frengen, J., Lindmo, T., Paus, E., Schmid, R., and Nustad, K., *J. Immunol. Meth.* **178**:141–151 (1995).
38. Martens, C., Bakker, A., Rodriguez, A., Mortensen, R. B., and Barrett, R. W., *Anal. Biochem.* **273**:20–31 (1999).
39. Horisberger, M., Rosset, J., and Bauer, H., *Experimentia* **31**:1147–1149 (1975).
40. Faulk, W. P. and Taylor, G. M., *Immunochemistry* **8**:1081 (1971).
41. Roth, J., *Histochem. Cell Biol.* **106**:1–8 (1996).
42. Horisberger, M. and Rosset, J., *J. Histochem. Cytochem.* **25**:295–305 (1977).
43. Perrin, A., Lanet, V., and Theretz, A., *Langmuir* **13**:2557–2563 (1997).
44. Perrin, A., Theretz, A., and Mandrand, B., *Anal. Biochem.* **256**:200–206 (1998).
45. Perrin, A., Theretz, A., Lanet, V., Vialle, S., and Mandrand, B., *J. Immunol. Meth.* **224**:77–87 (1999).
46. Mazzola, L. T. and Fodor, S. P. A., *Biophys. J.* **68**:1653–1660 (1995).
47. Olk, C. H., Heremans, J., Lee, P. S., Dziedzic, D., and Sargent, N. E., *J. Vacuum Sci. Technol. B* **9**:1268 (1991).
48. Quist, A. P., Bergman, A. A., Reimann, C. T., Oscarsson, S. O., and Sundqvist, B. U. R., *Scan. Microsc.* **9**:395 (1995).
49. Wadu-Mesthrige, K., Xu, S., Amro, N. A., and Liu, G.-Y., *Langmuir* **15**: 8580–8583 (1999).
50. Delamarche, E., Bernard, A., Schmid, H., Bietsch, A., Michel, B., and Biebuyck, H., *J. Am. Chem. Soc.* **120**:500–508 (1998).
51. Bernard, A., Delamarche, E., Schmid, H., Michel, B., Bosshard, H. R., and Biebuyck, H., *J. Am. Chem. Soc.* **14**, 2225–2229 (1998).
52. Jones, V. W., Kenseth, J. R., Porter, M. D., Mosher, C. L., and Henderson, E., *Anal. Chem.* **70**:1233–1241 (1998).
53. O'Brien, J. C., Jones, V. W., Porter, M. D., Mosher, C. L., and Henderson, E., *Anal. Chem.* **72**:703–710 (2000).
54. Kricka, L. J., in *Immunoassay*, E. P. Diamandis and T. K. Christopoulos (eds.), Academic Press, London, 1996, pp. 389–404.
55. Bottomley, L. A., *Anal. Chem.* **70**:425R-475R (1998).
56. Takano, H., Kenseth, J. R., Wong, S. S., O'Brien, J. C., and Porter, M. D., *Chem. Rev.* **99**:2845–2890 (1999).
57. Masai, J., Sorin, T., and Kondo, S., *J. Vacuum Sci. Technol. A* **8**:713 (1990).
58. Mulhern, P. J., Blackford, B. L., Jericho, M. H., Southam, G., and Beveridge, T. J., *Ultramicroscopy* **42–44**:1214 (1992).
59. Kawashima, K., Sisido, M., and Ichimura, K., *Chem. Lett.* 491 (1995).
60. Feng, C. -D., Ming, Y. -D., Hesketh, P. J., Gendel, S. M., and Stetter, J. R., *Sensors Actuators B: Chem.* **35**:431 (1996).
61. Eppell, S. J., Simmons, S. R., Albrecht, R. M., and Marchant, R. E., *Biophys. J.* **68**:671 (1995).
62. Putman, C. A. J., Grooth, B. G. d., Hansma, P. K., Hulst, N. F. v., and Greve, J., *Ultramicroscopy* **48**:177–182 (1993).
63. Schneider, S., Folprecht, G., Krohne, G., and Oberleithner, H., *Pfluger's Archiv.* **430**:795–801 (1995).

64. Smith, P. R., Bradford, A. L., Schneider, S., Benos, D. J., and Geibel, J. P., *Am. J. Physiol.* **272**:C1295–C1298 (1997).

65. Shaiu, W. L., Vesenka, J., Jondle, D., Henderson, E., and Larson, D. D., *J. Vacuum Sci. Technol. A* **11**:820–823 (1993).

66. Thimonier, J., Montixi, C., Chauvin, J.-P., He, H. T., Rocca-Serra, J., and Barbet, J., *Biophys. J.* **73**:1627–1632 (1997).

67. Nakano, K., Taira, H., Maeda, M., and Takagi, M., *Anal. Sci.* **9**:133 (1993).

68. Wagner, P., Hegner, M., Kernen, P., Zaugg, F., and Semenza, G., *Biophys. J.* **70**:2052 (1996).

69. Reed, L. J. and Muench, H., *Am. J. Hygiene* **27**:493–497 (1938).

70. Wagner, P., Hegner, M., Guntherodt, H.-J., and Semenza, G., *Langmuir* **11**:3867 (1995).

71. Hegner, B., Wagner, P., and Semenza, G., *Surf. Sci.* **291**:39 (1993).

72. Kumar, A. and Whitesides, G. M., *Appl. Phys. Lett.* **63**:2002–2004 (1993).

73. Chen, C. S., Mrksich, M., Huang, S., Whitesides, G. M., and Ingber, D. E., *Biotechnol. Progress* **14**:356–363 (1998).

74. Libioulle, L., Bietsch, A., Schmid, H., Michel, B., and Delamarche, E., *Langmuir* **15**:300–304 (1999).

75. Diamandis, E. P. and Christopoulos, T. K. (eds.), *Immunoassay*, Academic Press, London, 1996.

76. Wagner, P., Hegner, M., Kernene, P., Zaugg, F., and Semenza, G., *Biophys. J.* **70**:2052–2066 (1996).

77. Duhachek, S. D., Kenseth, J. R., Casale, G. P., Small, G. J., Porter, M. D., and Jankowiak, R., *Anal. Chem.* **72**:3709–3716 (2000).

78. Jones, V. W., Kenseth, J. R., and Porter, M. D., *Anal. Chem.* **70**:1233–1241 (1998).

79. Burgess, J. D., Jones, V. W., Porter, M. D., Rhoten, M. C., and Hawkridge, F. M., *Langmuir* **14**:6628–6631 (1998).

80. Finot, M. O. and McDermott, M. T., *J. Am. Chem. Soc.* **119**:8564–8565 (1997).

81. Nagao, E. and Dvorak, J. A., *Biophys. J.* **76**:3289–3297 (1999).

82. Sarma, V. R., Silverton, E. W., Davies, D. R., and Terry, W. D., *J. Biol. Chem.* **246**:3753 (1971).

83. Delamarche, E., Sundarababu, G., Biebuyck, H., Michel, B., Gerber, C., Sigrist, H., Wolf, H., Ringsdorf, H., Xanthopoulos, N., and Mathieu, H. J., *Langmuir* **12**:1997–2006 (1996).

84. Roberts, C. J., Williams, P. M., Davies, J., Dawkes, A. C., Sefton, J., Edwards, J. C., Haymes, A. G., Bestwick, C., Davies, M. C., and Tendler, S. J. B., *Langmuir* **11**:1822–1826 (1995).

85. Droz, E., Taborelli, M., Descouts, P., Wells, T. N. C., and Werlen, R. C., *J. Vacuum Sci. Technol. B* **14**:1422 (1996).

86. Grubor, N. M., Shinar, R., Jankowiak, R., Porter, M. D., and Small, G. J., *Biosens. Bioelectron.* **19**:547–556 (2004).

87. Kenseth, J., Wong, S.-S., Takano, H., Jones, V., and Porter, M., "Formation, Structural Characterization, and Reactivity of Organosulfur Monolayers on

Gold," *Fourier Transform Spectroscopy 12th International Conference*, Structures and Dynamics in Organic and Polymeric Systems Symposium, Tokyo, Japan, August 1999, K. Itoh and M. Tasumi, (eds.), Waseda University Press, 179–182.

88. Murphy, F. A., Gibbs, E. P. J., Horzinek, M. C., and Studdert, M. J., *Veterinary Virology*, 3rd ed., Academic Press, San Diego, 1999.
89. Bidawid, S., Malik, N., Adegbunrin, O., Sattar, S. A., and Farber, J. M., *J. Virol. Meth.* **107**:163–167 (2003).
90. Atmar, R. L. and Estes, M. K., *Clin. Microbiol. Rev.* **14**:15–37 (2001).
91. Chomel, B. B., *J. Vet. Med. Educ.* **30**:145–147 (2003).
92. Grabar, K. C., Brown, K. R., Keating, C. D., Stranick, S. J., Tang, S.-L., and Natan, M. J., *Anal. Chem.* **69**:471–477 (1998).
93. Mulvaney, P. and Giersig, M., *J. Chem. Soc. Faraday Trans.* **92**:3137–3143 (1996).
94. Vesenka, J., Manne, S., Giberson, R., Marsh, T., and Henderson, E., *Biophys. J.* **65**:992–997 (1993).
95. Donaldson, K. A., Kramer, M. F., and Lim, D. V., *Biosens. Bioelectron.* 20:322–327 (2004).
96. Ilic, B., Yang, Y., and Craighead, H. G., *Appl. Phys. Lett.* **85**:2604–2606 (2004).
97. Browning-Kelley, M. E., Wadu-Mesthrige, K., Hari, V., and Liu, G. Y., *Langmuir* **13**:343 (1997).
98. Park, K., Simmons, S. R., and Albrecht, R. M., *Scan. Microsc.* **1**:339–350 (1987).
99. Johnstone, R. W., Andrew, S. M., Hogarth, M. P., Pietersz, G. A., and McKenzie, I. F. C., *Mol. Immunol.* **27**:327–333 (1990).
100. Glaser, R. W., *Anal. Biochem.* **213**:152–161 (1993).
101. Takano, H., Kenseth, J. R., Wong, S., and O'Brien, J. C. P., *Chem. Rev.* **99**:2845–2890 (1999).
102. Croft, D., Shedd, G., and Devasia, S., *J. Dyn. Syst. Meas. Control* **123**:35 (2001).
103. Perez., H., Zou, Q., Devasia, S., *J. Dyn. Syst. Meas. Control* **126**:187 (2004).
104. Heller, M. J., Forster, A. H., and Tu, E., *Electrophoresis* **21**:157–164 (2000).
105. Ewalt, K. L., Haigis, R. W., Rooney, R., Ackley, D., and Krihak, M., *Anal. Biochem.* **289**:162–172 (2001).
106. Pankhurst, Q. A., Connolly, J., Jones, S. K., and Dobson, J., *J. Phys. D: Appl. Phys.* R167–R181 (2003).
107. Pale-Grosdemange, C., Simon, E. S., Prime, K. L., and Whitesides, G. M., *J. Am. Chem. Soc.* **113**:12–20 (1991).
108. Dong, Y. and Shannon, C., *Anal. Chem.* **72**:2371–2376 (2000).
109. Chang, S.-S., Shih, C.-W., Chen, C.-D., Lai, W.-C., and Wang, C. R. C., *Langmuir* **15**:701–709 (1997).
110. Wei, G.-T., Liu, F.-K., and Wang, C. R. C., *Anal. Chem.* **71**:2085–2091 (1999).

CHAPTER 5

Informatics Methods for Combinatorial Materials Science

CHANGWON SUH and KRISHNA RAJAN

Department of Materials Science Engineering
Combinatorial Materials Science and Materials Informatics Laboratory
Iowa State University
Ames, Iowa

BRANDON M. VOGEL

Polymers Division
National Institute of Standards and Technology (NIST)
Gaithersburg, Maryland

BALAJI NARASIMHAN and SURYA K. MALLAPRAGADA

Institute for Combinatorial Discovery
Department of Chemical and Biological Engineering
Iowa State University
Ames, Iowa

5.1. INTRODUCTION

Using modern combinatorial techniques to develop large materials databases is rapidly becoming essential in advance materials design. In order to interpret the data produced in large experiments in a useful fashion, it is necessary to assess not only one attribute from a combinatorial experiment but also multiple attributes that collectively contribute to the final "signal" of a material characteristic. A good example is spectral screening (such as chemical spectroscopy and diffraction) of combinatorial compositional libraries. While it is typical to identify a desired property that appears in a particular combinatorial array, a more challenging problem is to understand and monitor what changes are occurring among the various materials length scales. While the sophistication in screening property information in combinatorial experiment has evolved over the years, there are still severe bottlenecks in characterizing

Combinatorial Materials Science, Edited by Balaji Narasimhan, Surya K. Mallapragada, and Marc D. Porter

structural information in a detailed fashion from combinatorial compositional arrays. In this chapter we provide an introduction to ways to address this problem by integrating informatics techniques to "screen" high-throughput screening data and identify critical parts of a spectrum that may be offering insights into trends in the evolution of molecular structure and how it maps into the trends in properties in compositional libraries. We will present this discussion in the context of a case study on FTIR screening of combinatorial polymer libraries.

High-throughput experimentation (HTE) is essential to handle many parallel experiments by combinatorial synthesis and analysis techniques, and it leads to accelerate new drug discovery and screening of libraries. However, although these techniques provide a means of generating large amounts of experimental data, we still need to have a tool to interpret the data produced in HTE. A spectrum contains information on a variety of length scales of the material from which it is taken. Classical spectroscopy addresses the issue in deconvoluting the variety of contributions to the position, intensity, width, and/or shape of the peaks in a spectrum. This data analysis is supported through detailed simulations and comparisons with physical models to best fit a theoretical framework to allow for interpretation of the spectra. Hence the informatics challenge is to understand in a rapid manner the correlations between the multivariate contributions to a spectrum. The problem is really one of how to find correlations in high-dimensional datasets. In the next section we will summarize the mathematical approaches one may use to address this problem.

A first step in this process is to digitize the spectra as discrete datasets in order reduce the dimensionality of data such that we can begin to make subsequent data-mining tools to seek patterns of behavior and make predictions as well. There are many mathematical approaches to reduce the dimensionality of data; however, we will focus our attention on some basic linear methods known as principal component analysis

5.2. MATHEMATICAL STRATEGY FOR DIMENSIONALITY REDUCTION

Mathematically, principal-component analysis (PCA) relies on the fact that most of the descriptors are interrelated and these correlations in some instances are high. It results in a rotation of the coordinate system in such a way that the axes show a maximum of variation (covariance) along their directions. This description can be mathematically condensed to a so-called eigenvalue problem. The data manipulation involves decomposition of the data matrix **X** into two matrices, **V** and **U**. The two matrices **V** and **U** are orthogonal. The matrix **V** is usually called the "loadings" matrix, and the matrix **U** is called the "scores" matrix. The loadings can be understood as the weights for each original variable when calculating the principal component. The matrix **U** contains the original data in a rotated coordinate system. The mathematical analysis involves finding these new "data" matrices *U* and *V*. The dimensions of *U* (i.e., its rank) that captures all the information of the entire dataset of *X* (i.e., the number of variables) is far less than that of *X* (ideally 2 or 3). One now compresses the *N*-dimensional plot of the data matrix *X* into a two- or three-dimensional plot of *U* and *V*.

The eigenvectors of the covariance matrix constitute the principal components. The corresponding eigenvalues give a hint to how much "information" is contained in the individual components. The first principal component accounts for the maximum variance (eigenvalue) in the original dataset. The second principal component is orthogonal (uncorrelated) to the first and accounts for most of the remaining variance. A new row space are constructed in which to plot the data, where the axes represent the weighted linear combinations of the variables affecting the data. Each of these linear combinations is independent of each other and hence orthogonal. The data when plotted in this new space are essentially a correlation plot, where the position of each data point not only captures all the influences of the variables on that data but also their relative influence compared to the other data. Thus the *m*th PC (principal component) is orthogonal to all others and has the *m*th largest variance in the set of PCs. Once the *N* PCs have been calculated using eigenvalue/eigenvector matrix operations, only PCs with variances above a critical level are retained (screen test—described in the Appendix). The *M*-dimensional principal-component space has retained most of the information from the initial *N*-dimensional descriptor space, by projecting it into orthogonal axes of high variance. The complex tasks of prediction or classification are made easier in this compressed, reduced dimensional space. If we assume that information from the data can be gained only if the variation along an axis is a maximum, we have to find the directions of maximum variation in the data. In addition, these new axes should again be orthogonal to each other. In order to find the new axes, the direction of the maximum variation should be found first in order to take it for the first axis. Thereafter we use another axis that is normal to the first and rotate it around the first axis until the variation along the new axis is a maximum. Then we add a third axis, again orthogonal to the other two and in the direction of the maximum remaining variation, and so on (ideally we can capture all the meaningful relationships in 3 dimensions) [1,2].

Here, PCA is a data interpretation tool as a part of spectral informatics (Fig. 5.1). PCA is used to make new variables (PC) from linear combinations of existing variables. This chapter aims to describe the utility of PCA as a statistical modeling tool capable of accurate representation of complex, nonlinear, IR data without the need for a separate determination of the parameters typically required for conventional modeling.

By this method, data are utilized as a rotational way to find more useful information of data on the new sample and variable space. For the reason of

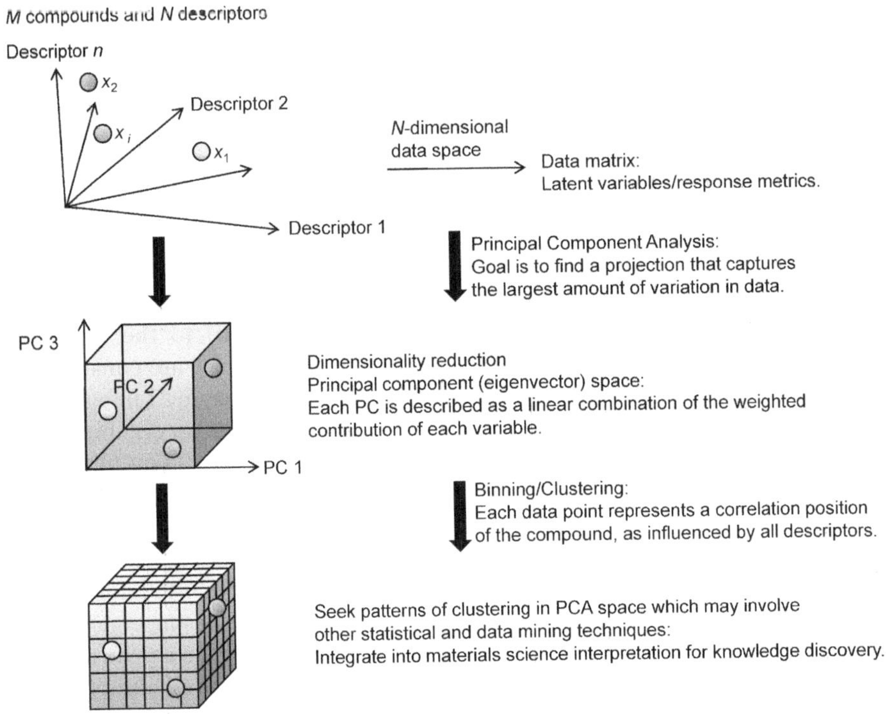

Figure 5.1. Schematic summarizing the procedural logic of principal-component analysis. [From Bajorath, J., Integration of virtual and high throughput screening, *Nature Rev. Drug Discov.* **1**:882–894 (2002); Rajan, K., Materials informatics, *Mater. Today* **8**:38–45 (Oct. 2005).]

convenience in handling spectral data, PCA has been applied to various spectral data such as Raman spectra [3], electron paramagnetic resonance (EPR) [4], and chromatography [5]. It was also used to detect minute chemical signals in the mixture spectra [6,7] or to determine the glass transition temperatures of polymeric thin films by identifying the highest peaks in the score plot of FTIR spectra [8].

5.3. CASE STUDY

Polyanhydrides are important biomaterials for biomedical applications, especially in drug delivery, due to their characters of degradation and erosion

[9–12]. For drug delivery applications, the choice of correct chemistry and management of various degradation factors (for example, kinetics of drug release and polymer-drug interaction) are essential for proper drug stabilization. To this end, Vogel *et al.* demonstrated high throughput methods to generate and screen compositional libraries of polyanhydride copolymers [8].

Figure 5.2 shows a Set of FTIR spectra and taking the digitized data, we can then set up PCA providing plots such as shown in Figure 5.3 (see also Figs. 5.4 and 5.5).

The apparent complexity of these PCA trajectories actually can be explained as follows. The PCA score plots are mapping the correlation between peaks. Consider a comparison between two peaks of the same wavenumber for two different chemistries. Assuming a change in peak intensity and peak position, we need to map the correlations in those features between the different peaks changes. The plot is actually a trajectory as it tracks the intensities in each of the spectra as one progressively goes through different wave numbers where

1. The axes represent intensities of peaks from two chemistries c1 and c2.

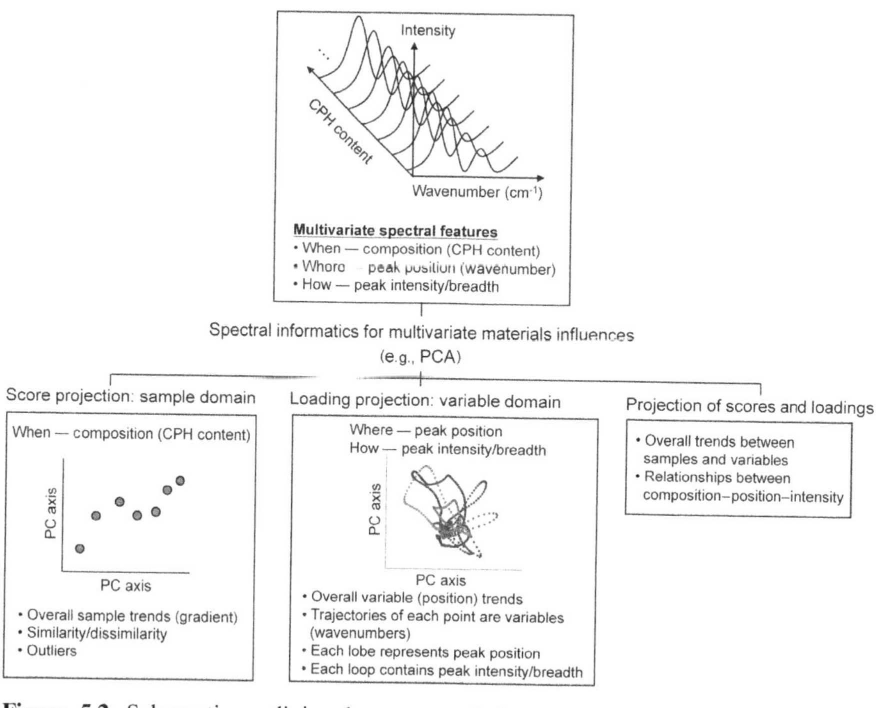

Figure 5.2. Schematic outlining how spectral data can be treated with principal-component analysis.

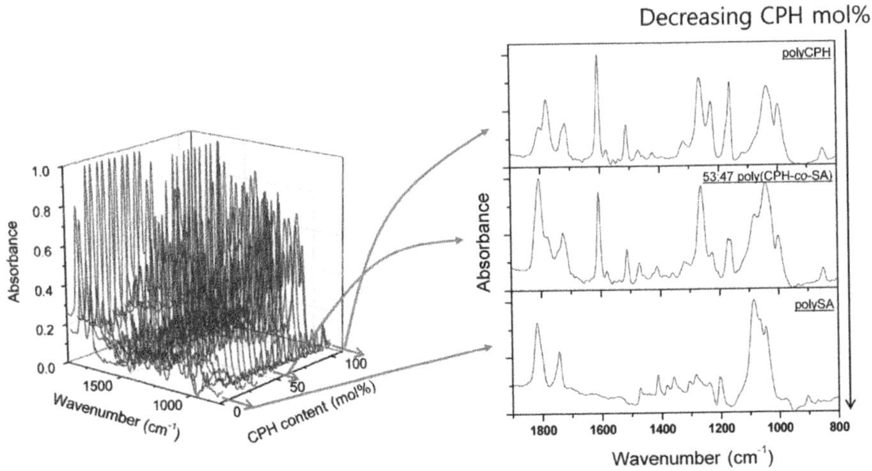

Figure 5.3. High-throughput FTIR spectra of combinatorial polyanhydride libraries with systematically varying compositions of poly(SA) and poly(CPH).

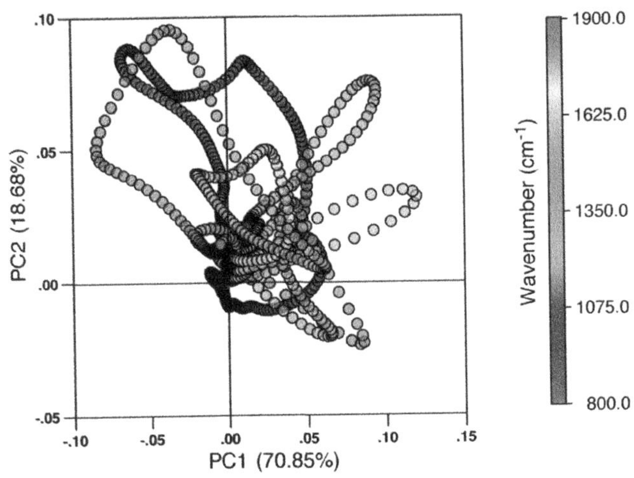

Figure 5.4. A PCA loading plot of the FTIR data shown in Figure 5.3. Each lobe is associated with a given wavenumber. The trajectory describes how each component of the spectral profile for a given wavenumber changes with respect to all the other intensities. The radial distance from the origin is associated with the magnitude of the intensity. These changes can be directly tracked to the compositional targets in the combinatorial arrays.

2. Two of the coordinates have to be (0, 0) since both peaks start at the background and return to the background.
3. The trajectory has to have two "maxima", one associated with each of the peaks.

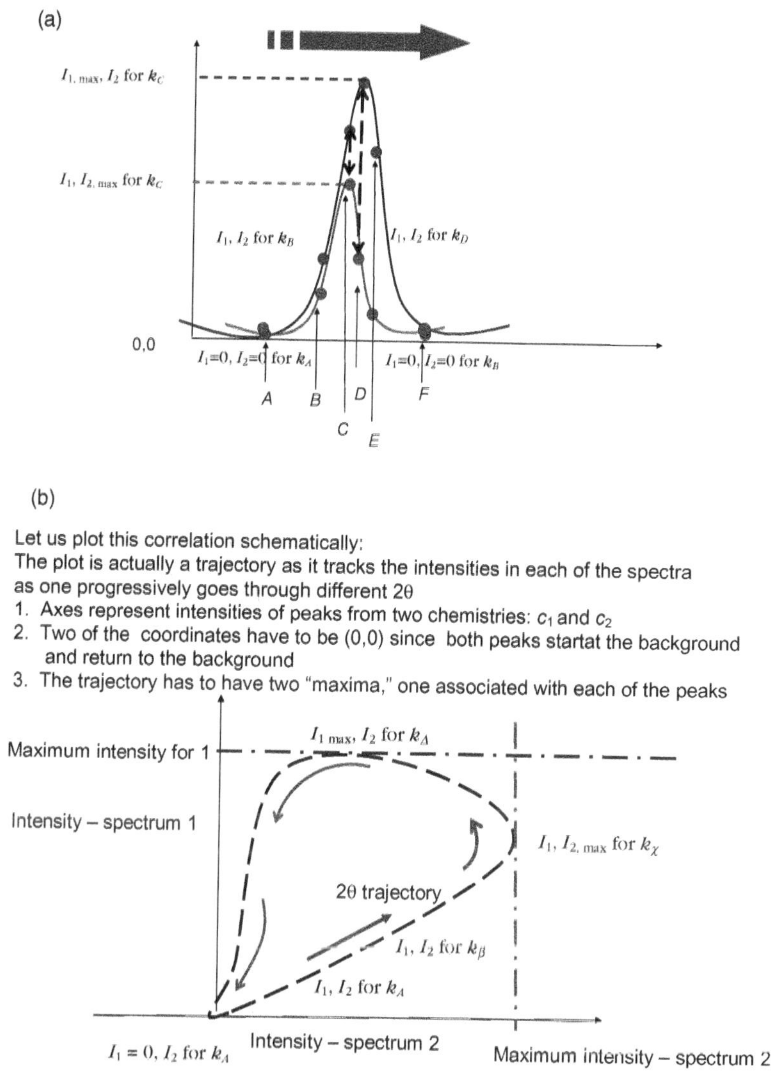

Figure 5.5. A schematic illustration of a bivariate problem of tracking correlations between spectral peaks. The PCA loading plot of Figure 5.4 is simply how an n-dimensional correlation plot appears but still retains many of the qualitative features shown in the bivariate case. The PCA loading plot is capturing the above logic for *all* the peaks *simultaneously* and not just two peaks contributing to the complex shapes of these loading plots.

If instead of monitoring the wavenumbers, we track the correlations according to the chemistry, then we can associate subtle changes in chemistry that cannot be seen by simply looking at the final output of the screening experiment. As a simple demonstration of that is to combine the score and loading plots together.

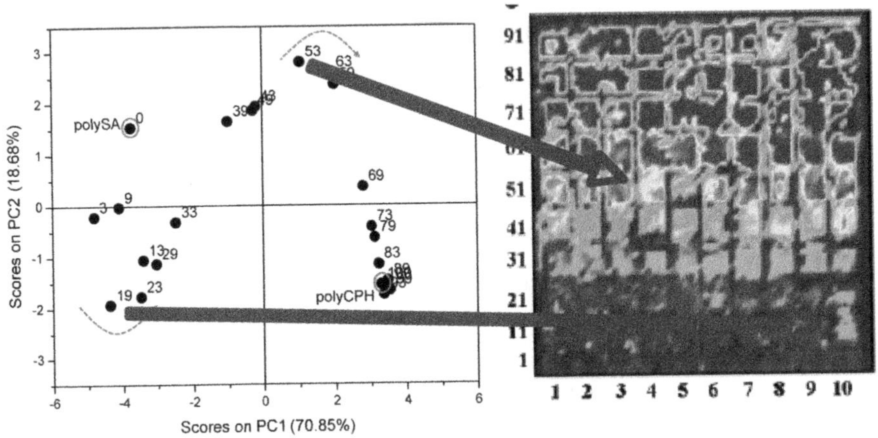

Figure 5.6. The PC1–PC2 score plot for 22 FTIR spectra. A score plot represents a sample domain and shows trends of each spectrum (sample). Two inflection points in score plots corresponding to compositions #19 and #53 are clearly seen on the projections of PC1–PC2. Numbers on figures represent compositions of CPH (e.g., for sample 3, it contains 3 mol% of CPH and 97 mol% of SA). Just by looking at the raw FTIR results from the combinatorial array, one could not have detected the fact the sample chemistries associated with #19 and #53 in the array are actually where significant changes in molecular structure influencing target properties are actually occurring.

Since score plots contain only information of each spectrum (samples), the use of scores and loadings elucidate the source of trajectories discussed in loading plots. In other words, when scores and loadings are considered simultaneously, the relationships between samples (spectra-CPH content) and variables (wavenumbers—peak positions) are detected. Thus, the first inflection point at #19 in Figure 5.6 shows that a region of 1610 cm^{-1} is burgeoning from the spectrum of #19 since this peak sits diagonally with sample #19 (e.g., the first quadrant—the third quadrant) as shown in Figure 5.7. As the CPH content is increased, a peak (at around 1610 cm^{-1}) becomes more dominant (i.e., experiences increased absorbance). Another inflection point at sample #53 on the score plot is due to a peak at around 1730 cm^{-1} (see Fig. 5.7). On studying those spectra in detail it was found that 19 : 81 poly(CPH-co-SA) and 53 : 47 poly(CPH-co-SA) were identified as critical spectra to change bonding characteristics of anhydrides. In other words, an aromatic–aromatic anhydride bond starts to be seen from 19 mol% of CPH. From 53 mol% of CPH, a peak assigned by an aliphatic–aromatic bond starts to affect the spectrum.

5.4. CONCLUSIONS

This example with FTIR data serves to demonstrate how simple visual inspection of combinatorial data is not sufficient in interpreting combinatorial

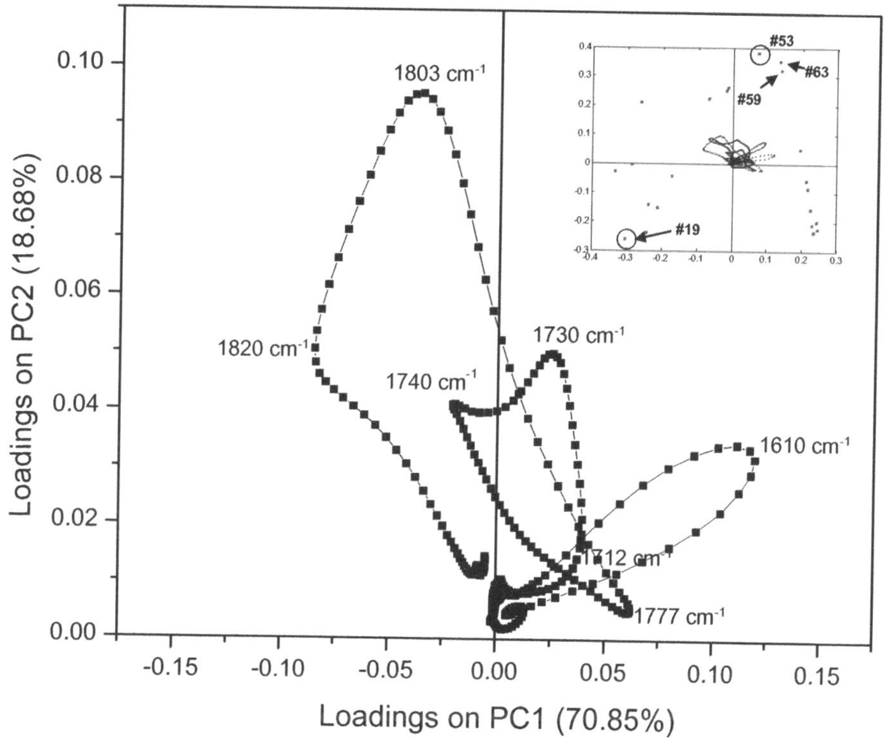

Figure 5.7. The PC1–PC2 loading plot for 22 FTIR spectra in the region 1900–1550 cm^{-1}. An inset is a biplot (i.e., score and loading plot on the same plane). A trajectory in the loading plot represents the change of wavenumbers. Each lobe is a peak position, and the size of each loop corresponds to intensity (absorbance).

data. The merging of informatics techniques with combinatorial experiments provides a significant "value added" level of interpretation that of combinatorial arrays. The use of such data-mining techniques needs to become an integral part of high-throughput screening methods in combinatorial experimentation.

ACKNOWLEDGMENTS

B. Narasimhan and K. Rajan acknowledge support from the Office of Naval Research (Award No. N00014-06-1-1176).

REFERENCES

1. Bajorath, J., Selected concepts and investigations in compound classification, molecular descriptor analysis, and virual screening, *J. Chem. Inform. Comput. Sci.* **41**:233–245 (2001).

2. Rajan, K., Materials informatics, *Mater. Today* **8**:38–45 (Oct. 2005).
3. Uy, D. and O'Neill, A. E., Principal component analysis of Raman spectra from phosphorus-poisoned automotive exhaust-gas catalysts, *J. Raman Spectrosc.* **36**:988–995 (2005).
4. Steinbock, O., Neumann, B., Cage, B., Saltiel, J., Müller, S. C., and Dalal, N. S., A demonstration of principal component analysis for EPR spectroscopy: Identifying pure component spectra from complex spectra, *Anal. Chem.* **69**:3708–3713 (1997).
5. Pate, M. E., Turner, M. K., Thornhill, N. F., and Titchener-Hooker, N. J., Pincipal component analysis of nonlinear chromatography, *Biotechnol. Progress* **20**:215–222 (2004).
6. Hasegawa, T., Detection of minute chemical species by principal-component analysis, *Anal. Chem.* **71**:3085–3091 (1999).
7. Shin, H. S., Lee, H., Jun, C., Jung, Y. M., and Kim, S. B., Transition temperatures and molecular structures of poly(methyl methacrylate) thin films by principal component analysis: Comparison of isotactic and syndiotactic poly(methyl methacrylate), Vibrat. *Spectrosc.* **37**:69–76 (2005).
8. Vogel, B. M., Cabral, J. T., Eidelman, N., Narasimhan, B., and Mallapragada, S. K., Parallel synthesis and high throughput dissolution testing of biodegradable polyanhydride copolymers, *J. Combin. Chem.* **7**:921–928 (2005).
9. Göpferich, A. and Tessmar, J., Polyanhydride degradation and erosion, *Adv. Drug Deliv. Rev.* **54**:911–931 (2002).
10. Kumar, N., Langer, R. S., and Domb, A. J., Polyanhydrides: An overview, *Adv. Drug Deliv. Rev.* **54**:889–910 (2002).
11. Mathiowitz, E., Kreitz, M., and Rekarek, K., Morphological characterization of bioerodible polymers. 2. Characterization of polyanhydrides by Fourier-transform infrared spectroscopy, *Macromolecules* **26**:6749–6755 (1993).
12. Eriksson, L., Johansson, E., Kettaneh-Wold, N., and Wold, S., *Multi- and Megavariate Data Analysis/Principles and Applications,* Umetrics Academy, Umetrics AB, Umeå, Sweden, 2001.
13. Wichern, D. and Johnson, R. A., *Applied Multivariate Statistical Analysis*, 5th ed., Prentice-Hall, Englewood Cliffs, NJ, 2002.
14. Geladi, P., Review: Cbemometrics in spectroscopy. Part 1. Classical chemometrics, *Spectrochim. Acta B* **58**:767–782 (2003).

APPENDIX: OPTIMAL NUMBER OF PRINCIPAL COMPONENTS

As discussed above, PCA is a method to decompose the data matrix X as

$$X = u_1 v_1^T + u_2 v_2^T + \cdots + u_F v_F^T + E_F = U_F V_F^T + E_F \tag{1}$$

Here $U_F V_F^T$ is the information of X and E_F is a noise. The optimum number F should be found while E_F is small. Since the eigenvalue represents variance, the percent variation explained by the corresponding principal component can be calculated from the respective eigenvalues. Thus

$$\frac{\lambda_1}{\text{Sum of all eigenvalues}} \times 100 \tag{2}$$

Therefore the percent variation explained by the first f components is

$$\frac{\lambda_1 + \cdots + \lambda_f}{\text{Sum of all eigenvalues}} \times 100 \tag{3}$$

Generally, F should be chosen to explain at least about 80–90% of variation of data in PCA. Kaiser's rule suggests choosing the number of eigenvalues larger than unity can be used to effectively choose the optimal number of principal components [12,13]. A screen plot which shows eigenvalues by each PC is useful to use Kaiser's rule.

CHAPTER 6

Combinatorial Approaches and Molecular Evolution of Homogeneous Catalysts

L. KEITH WOO
Institute for Combinatorial Discovery
Department of Chemistry
Iowa State University
Ames, IA

6.1. INTRODUCTION

This chapter is a brief survey rather than a comprehensive review of the use of combinatorial methods and molecular evolution in discovering and developing homogeneous catalysts. The reader is referred to several reviews and monographs for more detailed information [1–7]. The discussion that follows illustrates some of the general principles that guide this expansive field. Some specific cases of catalyst development are included to demonstrate representative approaches.

Catalysis is one of the most important applications in organic synthesis in terms of manufacturing bulk materials as well as producing fine chemicals. Although high yields have always been an important factor in chemical processing, eliminating unwanted side products is also a critical objective. Acute awareness of these goals has been raised by the concept of atom economy [8]. In ideal cases, all atoms in the reagents are used in forming only the product, with nothing left unused. In addition to conserving resources, processes that approach or reach atom economy also produce less waste and protect the environment. Catalysis can facilitate these aspects by increasing the selectivity for the desired product. Moreover, catalysts can promote reactions at lower temperatures and save energy in commercial applications. It is clear that catalyst development is crucial in the search for sustainable, green technologies.

Combinatorial Materials Science, Edited by Balaji Narasimhan, Surya K. Mallapragada, and Marc D. Porter

Catalysts can be broadly classified as homogeneous or heterogeneous. In the former case, all species are in the same phase. In heterogeneous systems, the reaction takes place at the surface of a solid catalyst. Advantages of heterogeneous catalysts include ease of separation of products from the reaction mixture, and robustness and reusability of catalyst. Combinatorial approaches for heterogeneous catalysis were recently reviewed [9]. These have advanced faster than those for homogenous systems since large, dense libraries of solids are accessible through conventional methods, including vapor deposition and sol-gel synthesis [5]. However, since molecular evolution involves homogeneous systems, this review will exclude further discussion of heterogeneous catalysts. Moreover, homogeneous catalysts have key advantages in their own right. For example, homogeneous catalysts are often well characterized and may involve only a single active site. This leads to higher selectivities and can facilitate mechanistic studies that lead to improved catalysts.

The growth of homogeneous catalysis is due largely to the preparation of fine and specialty chemicals through the use of soluble processes [10]. This includes the synthesis of pharmaceuticals and agrichemicals—bioactive organic compounds that typically have complex structures and often include stereogenic centers. Production of chiral molecules along with the need for high enantiomeric purity places a significant demand on chemical selectivity often best attained by homogeneous catalysis, ideally due to the involvement of only one type of active site. In addition to better regulation of the catalytically active species in homogeneous systems, better temperature and mixing control is achieved in the solution phase. Catalyst and reagent concentrations are also more readily controlled in homogeneous systems.

Traditional research in homogeneous catalysis has been devoted largely to the design of ligands and metal complexes with the goal of controlling the reactivity, selectivity, and stability of transition metal complexes. Approaches in ligand design include (1) knowledge of the mechanism, (2) experience and insight, (3) molecular modeling, and (4) serendipity. These strategies have led to the development of several successful catalytic processes. Representative of the importance of this traditional approach are recent Nobel Prizes in chemistry. Sharpless, Noyori, and Knowles were recognized in 2001 for developing homogeneous transition metal-based asymmetric catalysis. This pioneering work demonstrated that a chemical catalyst could operate with specificities previously seen only in biology [11,12]. The 2005 Chemistry Nobel Prize was awarded to Chauvin, Grubbs, and Schrock for olefin metathesis catalysis developed from organometallic complexes [13].

Despite recent advances, it is still exceedingly difficult to prepare a catalyst using first principles in a rational design approach. The collective interactions in the active site are often extremely complex, and it is not always clear how to specifically catalyze the transformation of one or more molecules into new compounds. Moreover, simple changes in substrate structure can often require tedious reoptimization of the catalyst as the interplay between the metal catalyst and reagents is highly dependent on the ligand environment. The complexities of many interdependent variables has forced the use of empirical methods to identify and optimize metal complexes for catalysis. Consequently, the entire catalyst development process is enormously time-consuming and requires several iterations of trial followed by modifications. Typically, a few candidates are made at a time and structure activity relationships are analyzed individually.

Subsequent fine tuning of the catalyst properties are implemented by modifying the environment around the central metal center (active site) through tailoring specific aspects of the ligand structure. The importance of ligand effects are aptly illustrated by nickel-catalyzed reactions of butadiene (Scheme 6.1). By changing the ligand on nickel catalysts, the product selectivity can be varied from linear dimers, cyclic dimers, cyclic trimers, polymers, and other compounds [14].

Chemical intuition has played an important role in ligand design. However, the application of this approach has its caveats. For example, Knowles and coworkers initially thought that chirality at the phosphorus atom of ligands used in asymmetric hydrogenation was a key factor in the observed enantioselectivity. This was subsequently proved wrong with the development of excellent catalysts bearing phosphine ligands that were chiral at carbon backbone positions [15].

Catalyst optimization also involves an understanding of the elementary pathways at the molecular level. This entails careful mechanistic studies to identify rate-limiting and product-forming steps. One can then attempt to modify the catalyst or conditions to enhance desired elementary steps. In addition, an understanding of decomposition routes may allow manipulation of the system to increase the lifetime of a catalyst. Computational chemistry has made tremendous contributions in catalyst design, but it must still be followed with experimental studies to optimize catalysts through trial and error. A key, unsolved grand challenge is correlating first principles with catalyst design.

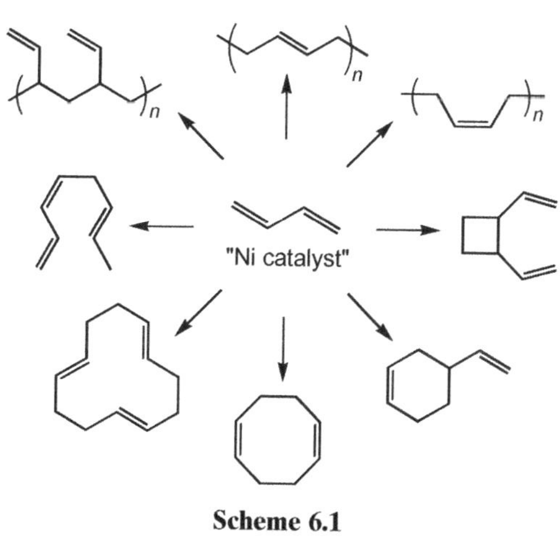

Scheme 6.1

The quintessential catalysts are enzymes. These remarkably selective and efficient chemical machines have evolved over billions of years. Much progress has been made in understanding, designing, and preparing synthetic peptides that adopt a desired folding pattern in solution. However, achieving enzyme-like activity in artifical proteins using rational design is still exceedingly difficult [16]. Biomimetic approaches have been utilized extensively in many laboratories for designing small molecule models of enzyme active sites [17,18]. This strategy has been extremely valuable in understanding biological processes as well as developing new catalytic systems. However, chemists' attempts at mimicking nature with model complexes have not had great success in producing useful catalysts. The first effective biomimetic catalysts were described in 1998 [19,20]. An alternate approach is utilizing wildtype enzymes for chemical transformations in the laboratory. The use of biocatalysis in the laboratory and in industry is growing [21–23]. Although thousands of enzymes are known, this is not always practical. Enzymes can be difficult to isolate and are not always readily applied in their natural state. Moreover, an enzyme is often limited to one substrate and/or a restricted range of reaction conditions.

Site-directed mutagenesis has been implemented to reengineer natural enzymes for use in a desired reaction, improve stability, and increase efficiency [24–27]. However, applying rational design to replace a specific amino acid has only provided modest success in protein engineering. The function of an enzyme is a result of the interplay of many amino acids participating in 3D space. Changes in amino acids that are either proximal or distal to the active site can significantly affect activity [28]. Many aspects of enzyme catalysis are still not understood. NMR and X-ray structural studies have provided a great deal of information, but only provide a static picture of a dynamic process. Even if some aspects of the reaction coordinate are hypothesized, it is still not possible to design an active site to achieve the necessary dynamic properties. Thus, point mutagenesis is perhaps more difficult and tedious than the rational design of ligands for homogeneous catalysis.

Catalytic science has made tremendous advances in developing efficient and selective catalysts. However, the rational design of catalysts still remains a difficult task and is still largely an art. The challenges in controlling chemical reactivity is magnified by the vast search space of possible structures and compounds which may lead to productive catalysts. However, traditional catalyst design is typically limited by classes of ligands and probable metal choices. Clearly, a high-throughput combinatorial strategy would expedite catalyst design and development.

6.2. COMBINATORIAL CATALYSIS

To meet the future needs for new materials, fine chemicals, therapeutics, etc., development of the next generation of catalysts will need to go beyond conventional practices. Traditional catalyst design is generally limited by conceivable ligands and foreseeable metal choices. Combinatorial methods expand the search to include both likely and unlikely combinations of metals and ligands. This approach elicits the possibility of finding a solution that traditional methods will ignore or have difficulty in finding [29]. The use of combinatorial strategies is a new paradigm in which catalyst development and optimization are accelerated by orders of magnitude over the traditional one-sample-at-a-time approach. Large numbers of diverse samples are created collectively and subsequently evaluated by high-throughput processes. The use of rapid screening techniques is swiftly growing as an expedient means of optimizing structure and function of new catalysts. In many cases, the screening is accomplished through robotics and/or highly parallel processing. Three general combinatorial approaches have been used in developing small-molecule homogeneous catalysts: (1) varying conditions or additives with known transition metal complexes, (2) preparing libraries of ligands to modify the environment of a transition metal ion, and (3) screening a library of substrates with a single transition metal complex. The first two strategies are most commonly employed and involve screening a library of catalysts with the same substrate or same pair of reaction partners. Although combinatorial approaches will not supplant traditional methods, they will be an important complement to rational catalyst design.

An example illustrating the combinatorial approach of varying additives was employed with a palladium-catalyzed annulation reaction of indoles [Eq. (6.1)]. In this case, arrayed microreactors were rapidly assayed by multiplexed capillary electrophoresis to survey 88 combinations for catalytic activity, selectivity, and kinetics. This provided a rapid means of finding an efficient palladium catalyst for the synthesis of carbolines (Fig. 6.1) [30].

(6.1)

The discovery of an exceptionally selective catalyst for the production of chiral hetero-Diels–Alder products [Eq. (6.2)] was achieved with a combinatorial library of titanium(4+) complexes based on chelating diol ligands (Scheme 6.2) [31]. Thirteen chiral diols (Scheme 6.3) were used in a parallel manner to produce a library of 104 catalysts. Catalysts composed of two diols with the composition $Ti(L_m)(L_n)$ ($m,n = 4$–7) were found to catalyze reaction (6.2) with yields up to 100% and 99% enantiomeric excess (ee).

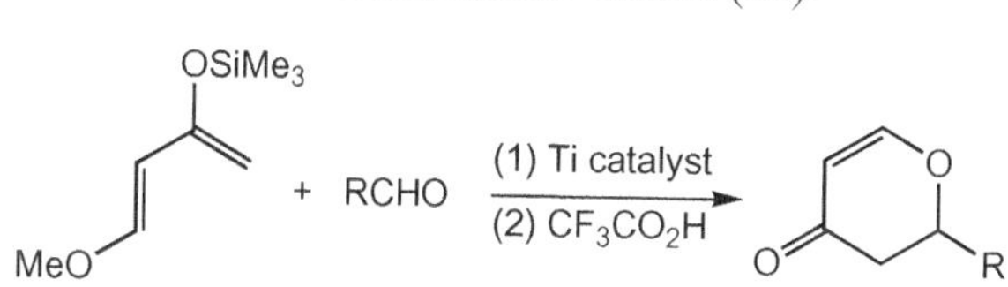

(6.2)

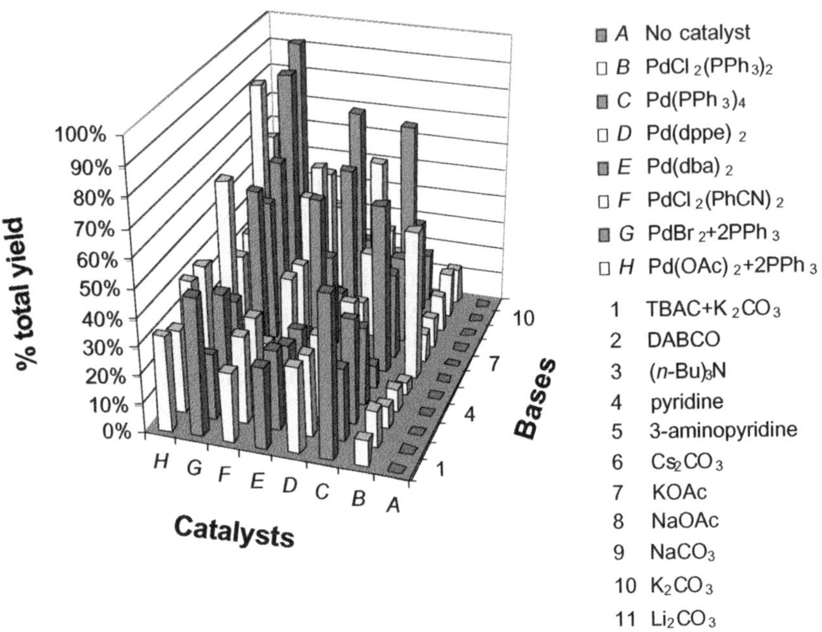

Figure 6.1. Results of a multiplexed chemical etch separation from 88 microreactors. (Reprinted by permission from Ref. 30. Copyright 2000 American Chemical Society.)

	L_0	L_1	L_2	• • •	L_n	• • •	L_{13}
L_0	L_0TiL_0	L_0TiL_1	L_0TiL_2				L_0TiL_{13}
L_1		L_1TiL_1	L_1TiL_2		• • •		L_1TiL_{13}
L_2			L_2TiL_2		• • •		L_2TiL_{13}
⋮							⋮
L_m					L_mTiL_n		⋮
⋮							⋮
L_{13}							$L_{13}TiL_{13}$

Scheme 6.2 (Reprinted by permission from Ref. 31. Copyright 2002 American Chemical Society.)

A major challenge in combinatorial approaches is to develop strategies for the synthesis of relevant, unbiased ligand libraries with high diversity. By analogy to proteins with 20 amino acids as building blocks, the most efficient means of preparing diverse libraries typically involve modular ligand synthe-

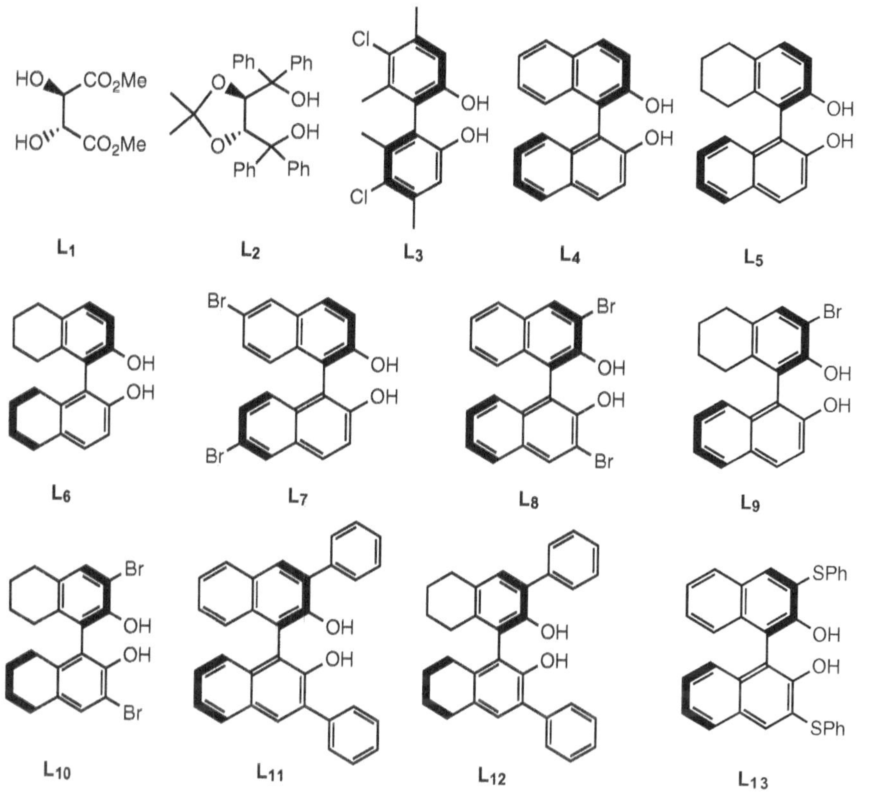

Scheme 6.3 (Reprinted by permission from Ref. 31. Copyright 2002 American Chemical Society.)

ses. The ideally designed screening library should contain compounds with a variety of structural shapes and molecular properties, while avoiding redundancies. Other requirements involve the need to find structurally and compositionally relevant samples with promising attributes that allow the library to be reduced to a smaller subset of lead compounds. Once a smaller set of lead compounds is identified for a more focused library, higher purity and better characterization can be achieved. Subsequently, more careful experimentation and optimization can be done on individual catalysts.

Two approaches are generally employed to prepare combinatorial libraries: (1) parallel synthesis and (2) split-and-pool (Fig. 6.2). Both types of library syntheses are also limited to the preparation of modular type ligands. Parallel methods were developed by Geysen [32], in which arrays of peptides were synthesized on pin-shaped solid supports and by Houghten [33], using a technique for creating peptide libraries in tiny mesh "tea bags" by solid-phase synthesis. This strategy involves creating sets of discrete compounds simultaneously in arrays of physically separate reaction vessels or microcompartments without interchange of intermediates during the assembly process. This

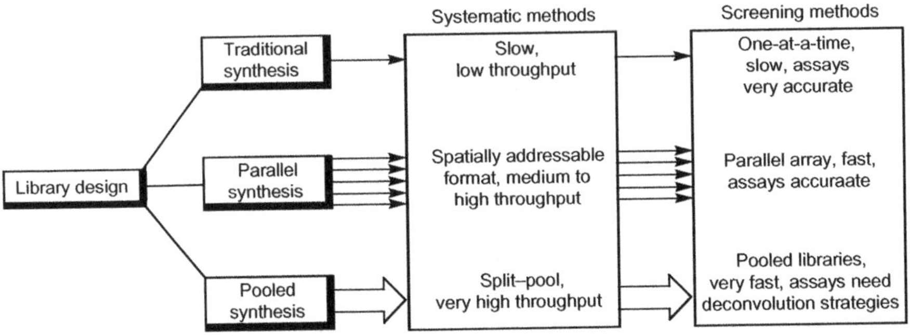

Figure 6.2. Traditional, parallel, and split–pool comparison. (Reprinted by permission from Ref. 1. Copyright 2002 Wiley-VCH.)

approach restricts the size of the library and its inherent diversity based on the limitations in the number of reactions that can be run in parallel. The advantage of the parallel approach is that each member of the library is in a specific spatial address or container. Addressable libraries allow each catalyst to be tested individually in highly parallel screening [5].

In the split/pool strategy, a solid support is divided into sets of individual samples and each set is subjected to a reaction with a different building block (Fig. 6.3). The samples are recombined (pooled) to produce a single batch of solid supports bearing a mixture of components. The pool is randomly redivided (split) and the new sample sets are again treated with a different building block. Repetition of these processes produces a library where a discrete particle of solid support carries a single library member, and the number of members is equal to the product of the number of building blocks times the number of reactions. In this manner, multiple substrates are transformed simultaneously under a common set of reaction conditions, producing all possible combinations of a given set of building blocks attached to a solid support in a minimum of steps. This approach produces large libraries that require effective deconvolution strategies [34] and can be difficult to screen.

Early combinatorial studies involve the development libraries based on analogs of known ligand types using a modular building block approach. These efforts were modest in scale, assaying 11 catalysts and two different substrates [36]. No superior catalysts were identified, but the application of combinatorial methods to find asymmetric catalysts was demonstrated. Subsequently, key examples have been reported that demonstrate the ability of combinatorial methods to find catalysts that are not easily identified by traditional one-at-a-time approaches [37].

Generating sufficient chemical diversity so that the search space is adequately represented is one of the key challenges in library design. It is well known that sheer numbers alone do not insure diversity in a library. Successful strategies require innovative reactions to prepare diverse catalysts [38]. This

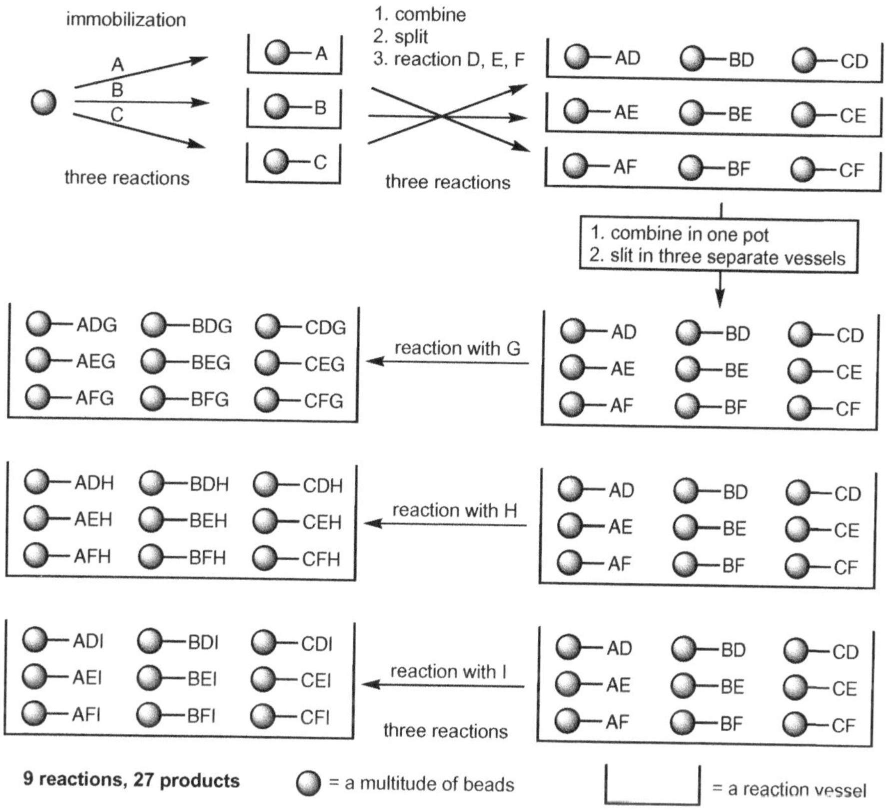

Figure 6.3. The split-and-pool approach. (Reprinted by permission from Ref. 35. Copyright 2002 Wiley-VCH.)

can lead to finding solutions that were not originally anticipated. Sensitivity, efficiency, and discriminating power of the screening method are also critical. Catalysts with the desired activity, if present, are often in extremely minute quantities in the library. The ability to access this small population of the pool is often a major obstacle. For large libraries such as those produced by the split and pool approach, data management, design of experiments, and informatics approaches are necessary to ensure that the library synthesis and high-throughput screening are executed properly and cross-validated. This includes identifying statistical noise such as artifacts, false positives, and false negatives.

Parallel methods involving modular synthesis have been reported for a number of classes of ligands and are the most widely used method for small-molecule catalysts [39]. The development of an efficient catalyst for the Strecker reaction [Eq.(6.3)] illustrates the power of this approach.

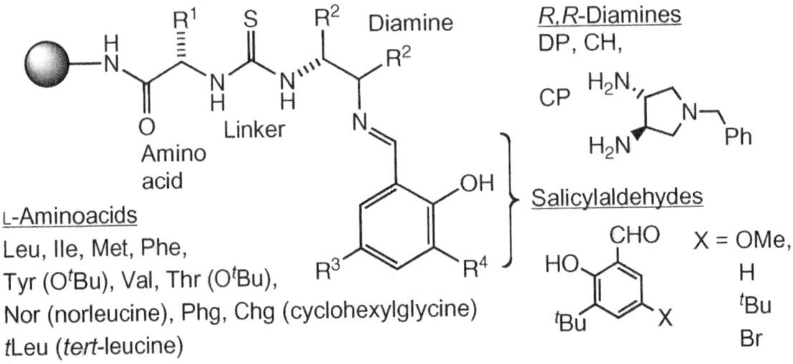

Scheme 6.4 (Reprinted by permission from Ref. 37. Copyright 1998 American Chemical Society.)

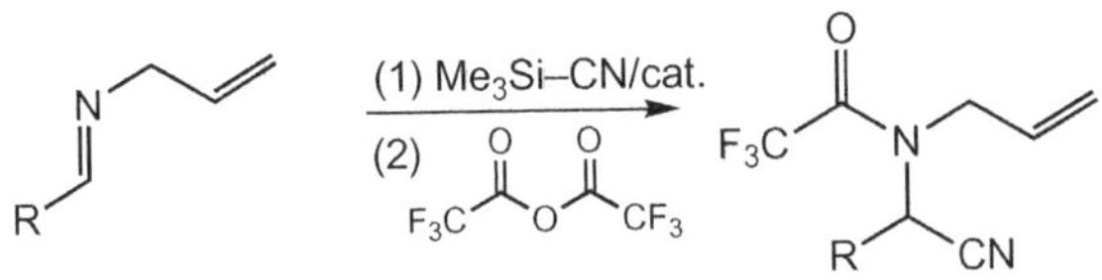

(6.3)

Various imine ligands were prepared from amino acids, diamines, and salicylaldehyde building blocks to form a library of 132 catalysts (Scheme 6.4) [37]. The best catalyst was found to have a selectivity of 91% ee.

The preceding approaches have used chemical means to develop structure and function. Practical issues have limited the size of the libraries developed through parallel strategies and combinatorial approaches for homogeneous transition metal catalysts are far from mature. An additional consideration is that chemical libraries based on synthetic structures cover a vastly different area of diversity space compared to combinatorial libraries constructed from biological building blocks [40]. Moreover, nature has at its disposal biological systems that are inherently combinatorial. Discussed next are biological systems that can be exploited in generating large libraries and methods for screening or selecting for function.

6.3. CATALYTIC ANTIBODIES

The search for selective catalysts has led to increased use of enzymes in synthesis. Although some 4000 enzymes have been identified [4], many useful chemical transformations occur that are not catalyzed by known enzymes. In order to harness the potential for engineering enzymes, powerful techniques such as catalytic antibodies, phage display, and directed evolution are employed.

Antibodies are a class of proteins elicited in an immune response to bind foreign species in the early stages of fighting disease. However, unlike enzymes, natural antibodies only function by selective binding. Catalytic antibody technology utilizes the immune process to reprogram antibodies to function as enzymes. This approach merges the combinatorial diversity of the immune system with a judiciously planned design by the scientist [4]. Any substance that triggers the creation of antibodies in the immune system is called an *antigen*. During an immunization response, a diverse library of antibodies is produced as a first step in the protection against foreign invaders. A first generation antibody from this library is selected on the basis of its binding ability for the antigens and the DNA for the variable region of this "lead" antibody is altered to create a new diversified library. Thus, the affinity of an antibody for a foreign substance is enhanced by somatic mutation.

In the laboratory, antibodies can be raised specifically to one target compound. If the foreign species is a small molecule (termed a *hapten*) that contains chemical information for a reaction, it should be possible to make an enzyme from antibodies. The basis for this concept originated from Pauling's deduction that the catalytic ability of enzymes reflected their complementarity to the transition state of the reaction of interest [41]. The ability of enzymes to align reactants for a specific transformation is a key attribute in their efficiency. The seminal suggestion of using antibodies to produce new catalysts using transition state analogs was proposed by Jencks in 1969 [42]. Antibodies produced in this manner have been shown to catalyze desired transformations [43,44]. Moreover, the immune system illustrates that natural evolution does not need to occur on the geologic timescale, but may be done within a timeframe of weeks. Thus, to shorten the evolutionary timeframe so that catalysts can be generated on the laboratory timescale, chemical instructions must be applied to the system.

The key concept in developing catalytic antibodies is to raise them to analogs of the transition state, using the ability of antibodies to selectively bind species ranging from small molecules to macromolecules with high affinity. In addition, structural studies show that antibodies have binding sites that are similar in dimension to the actives sites of enzymes [45,46]. When the hapten is a molecular analog of the transition state for a reaction, it serves as a structural template that produces a binding pocket within the antibody that mimics an enzyme active site. Construction of such a site should lead to catalysis, if complementarity of an active site to a transition state is a key factor in an enzyme's activity. The antibody should then divert part of its binding energy for the substrates toward inducing a geometry that resembles a transition state, accelerating the desired reaction and promoting catalysis. In other words, if a stable molecule that mimicks a fleeting transition state is used as an antigen, enzyme-like proteins should be obtained. This approach is now well established and has produced a variety of catalytic antibodies with rate accelerations exceeding factors of 10^5 over the uncatalyzed reaction. Catalytic antibodies have been produced for reactions ranging from acyl transfer to porphyrin metallation.

The catalytic antibody for porphyrin metallation, ferrochelatase antibody 7G12, exemplifies the principle of transition state complementarity. Ferrochelatase, the natural enzyme that catalyzes the insertion of iron(II) into protoporphyrin IX was found to be strongly inhibited by *N*-alkylporphyrins [47]. This inhibitor is highly distorted from the normally planar porphyrin structure due to repulsions of the *N*-alkyl substituent with other nitrogen atoms in the framework (Fig. 6.4) [48]. These data suggested that the natural enzyme catalyzes the metal ion insertion by distorting the protoporphyrin substrate into a conformation that makes the nitrogen lone pairs more available for the initial chelation step. Thus, using *N*-methylmesoporphyrin as a transition state analog produced antibody 7G12. This antibody was shown to be catatytic for insertion of metal ions into mesoporphyrin [Eq. (6.4)] with rates ($80\,h^{-1}$) that are comparable to those of ferrochelatase ($800\,h^{-1}$) [49]. The enzyme specificity constant for the antibody (k_{cat}/K_M) is $450\,M^{1}\,s^{-1}$.

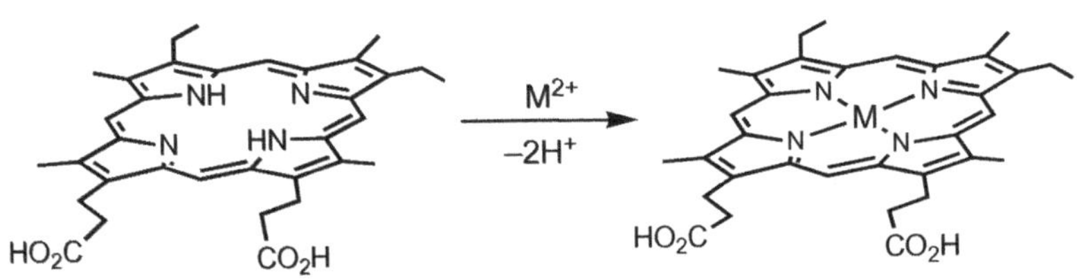

(6.4)

The primary antibody repertoire has been estimated to consist of 10^8 variants [50,51]. Other reports on the immune library, estimate that the diversity is as high as 10^{12} different antibodies [52]. In the primary method for eliciting antibody catalysts, an animal (typically, a mouse, rabbit, or chicken) is injected/immunized with a stable analog of the transition state for a given reaction. An immunization schedule over a period of weeks is used to generate a good antibody library [53]. The antibody-generating cells (B cells) are then harvested with a hybridoma technology [54]. This allows an antibody of a single specificity to be reproduced indefinitely in clones of the B cells. Thus, in principle, each monoclonal antibody in the immune repertoire library that was

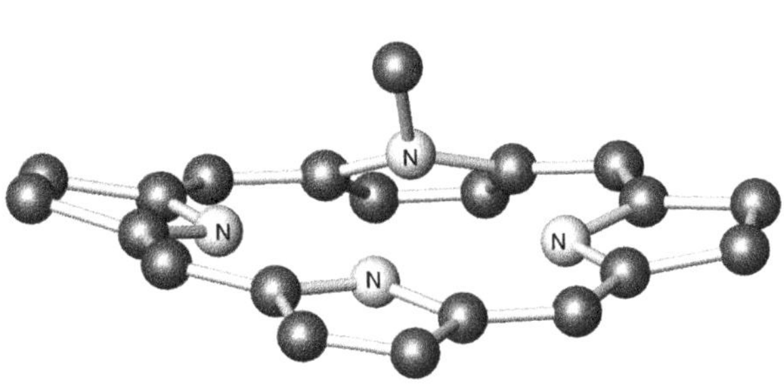

Figure 6.4. Molecular structure of the distorted core of an *N*-methylporphyrin. Hydrogen atoms and peripheral substituents are omitted. (Reprinted by permission from Ref. 48. Copyright 1982 American Chemical Society.)

elicited to the hapten is expressed in milligram to gram levels, isolated in high purity, and screened for catalytic activity. To date, more than 100 chemical reactions can be catalyzed by antibodies. Despite the success of this approach, practical applications are still rare, largely because most catalytic antibodies do not yet match the efficiencies of their natural counterparts [55,56].

A key challenge in developing catalytic antibodies is the design of a transition state analog (TSA). For each new reaction, the transition state analogue must be optimized to enhance the chances for identifying a catalytic antibody. Selectivities of antibody-catalyzed reactions reflect the structure of the TSA. Consequently, antibodies are able to discriminate between similar configurational or stereochemical isomers. In ideal cases, catalytic antibody performance could rival or exceed natural enzymes. Moreover, catalysis that cannot be achieved through traditional methods may be possible. However, by the very nature of being a ground-state molecule, the analog may not truly reflect the exact features of a short-lived transition state. Subtle differences between the mimic and the actual transition state can contribute to poorer performance of catalytic antibodies versus the corresponding natural enzyme. Thus, design principles focus on incorporating essential features of the transition state geometry.

Enzymes use a large repertoire of means to accelerate reactions. These include hydrophobicity, H-bonding, electrostatic potential, and general acid–base functions. Moreover, metal ions are also employed in many enzymes to extend their capabilities beyond that of the 20 natural amino acids. As analogues to metalloenzymes, semisynthetic catalytic antibodies have also been developed that are derivatized with a metal-ion-binding cofactor in the active site. This allows catalytic properties to be expanded to reactions such as oxidations [57]. An alternate method to the hapten design strategy is the "bait and switch" approach developed for raising antibodies to disfavored processes [58]. This method combines geometrical constraints with electrostatic stabilization of high-energy intermediates by accommodating both charged species and alignment of π clouds.

Nature uses a substantial variety of protein structures to produce enzymes [59,60]. However, the antibody immunoglobulin scaffold is not one naturally employed for catalysis. This may be an intrinsic limitation in the use of catalytic antibodies. In addition, no ground-state molecule can exactly model an actual transition state structure. Thus, selection of antibodies using inexact transition state analogs is unlikely to produce the ideal synthetic enzyme active site for catalysis. Furthermore, transition state analogues many not generate the best catalyst as other parameters are involved in biocatalysis [61]. Moreover, catalytic antibody technology involves only a single evolutionary round. The selected antibodies cannot be reintroduced into the immune system for further cycles of refinement. In addition, selection and mutagenesis of an antibody library is not a trivial task [62]. Although antibodies are often more primitive than native enzymes, their simplicity can have advantages in providing insights into catalytic function. Nonetheless, this technology has progressed from simple model reactions to complex processes. New challenges lie ahead such as delivering new catalytic processes for which natural enzymes do not exist or for reactions that are difficult to promote using existing methods.

Phage display of antibodies has been demonstrated. This has the ability to bind antigens and mimic the immune system to produce the desired binding molecules [63]. The application of in vitro methods of evolution to phage libraries, allows screening for catalytic properties. This topic is discussed in Section 6.4.2.

6.4. IN VITRO EVOLUTION METHODS

Inspiration for in vitro evolution is derived from nature, which uses the sequential steps of variation (diversity), selection, and amplification. The subsequent iteration of this sequence enables evolution to occur. Over billions of years, organisms have evolved in nature by repeated cycles of mutation, recombination, and survival of the fittest (selection). This process is widely known as Darwinian evolution. In an artificial but controlled manner, evolution has been used by animal and plant breeders for centuries. In practical applications of breeding, the selection process is achieved by culling offspring with undesired traits. Thus, classical breeding has been used to elicit desirable traits in whole organisms, plants and animals. In this process, the entire genome is optimized.

Molecular biology now allows us to elicit analogous improvements in shorter genetic sequences. This was first demonstrated in the laboratory in the late 1960s [64,65]. The terms "molecular breeding" [66] has been coined to describe this process. "Directed evolution" is also frequently used terminology for this strategy. In all cases, the developing population must have its genotype (inheritable information) linked to its phenotype (physical manifestation of functional properties) to achieve Darwinian evolution. By using molecular evolution, an initially identified, albeit crude, catalyst can be evolved and improved further by adding the element of mutation, combined with a specific selection pressure. Just as evolution in nature has produced sophisticated enzymes, in vitro evolution can be used in an analogous process in the lab. Thus, tailored catalyst performance can be developed in the manner of classical breeding, but on a molecular level.

In order to realize evolution, it is necessary to sequentially and repetitively create diversity and to apply a selective pressure [67]. The evolutionary process should provide the selected variant, in this case a desired catalyst, with a proliferation advantage. Each generation of catalysts becomes the new parents in a subsequent round of mutation and selection. The production of subsequent generations requires an ability to maintain a link between the genotype and phenotype so that the responsible encoding gene can be recovered. In biological systems, this is readily accomplished because the genome and proteome are both contained within the same cell or organism. Examples of how this is achieved for in vitro methods are discussed in the following sections. In addition, amplification of the phenotype is also an important step after each selection as this allows efficient species that are present in small amounts to become dominant within the population.

Highly efficient enzymes not only catalyze a desired reaction but also have evolved to avoid undesirable pathways [68]. Thus, a small change in amino acid sequence could significantly enhance alternative reactions or produce a drastic change in delicately balanced chemical processes. Random point mutation is one means of increasing the diversity of the sample library (population). However, only a few examples demonstrate that a single point mutation leads to significant improvement in catalytic activity [69]. Holland has demonstrated that the best way to achieve diversity in a DNA sequence is to integrate a high recombination rate with a low mutation rate [70]. This enables several beneficial DNA sequences that were achieved separately to be combined. A variety of mutation methods have been developed and these are listed in Table 6.1. A completely random method of mutagenesis is necessary to maximize an effective search through sequence space.

Often the useful variants in a library are rare and present in minute quantities in a complex mixture of undesirable molecules. Thus, the ultimate success of directed evolution is highly dependant on the strategy used to identify the best mutant [71]. This can be the crucial and most difficult task that determines the outcome. Two general approaches for identifying the desired catalyst are screening and selection. Each has its own set of advantages and disadvantages. Screening strategies typically involve an active search of all members in a library using high throughput methods that are miniaturized and automated. Sensitive assays can be difficult to perform and typically need expensive instrumentation. Screening can be facilitated by incorporating cues such as the production of fluorescence upon product formation. Typical screening techniques can assay libraries with up to 10^5 samples [72].

The alternative to screening is selection. This method exploits situations whereby the desired species remains in the library and the inactive variants are eliminated during the selection. In biological systems, it may be possible to design the catalytic search such that the formation of product is necessary for continued existence of the host cell. However, enzymes evolved for the survival benefit of an organism may not exhibit features essential for in vitro applications [100]. Alternatively, the desired property must be recognized by a biological assay. In abiological systems, it may be possible to link the desired property with immobilization of the desired species. In this manner, the culling step replicates survival of the fittest associated with natural Darwinian evolution. If appropriate selection conditions can be incorporated into the assay, library sizes $>10^{10}$ can be analyzed. In vivo selection necessarily involves limited conditions and may be poorly controlled. Evolution involving in vivo cloning in cells theoretically could be performed indefinitely. However, host cells eventually propagate as well as the wildtype strain, and it becomes difficult to apply selection pressure for additional improvement. In vitro methods

TABLE 6.1. Different Recombination Techniques with Their Main Advantages and Disadvantages

Group	Technique	Members
Shuffling	Recombination of small fragments based on homology in the sequence between mutations that stem from all kinds of mutagenesis strategies or different family members; aims for high recombination, but difficult to separate close mutations	Shuffling [74,75] Family shuffling [76] RE cut shuffling [77] ssDNA shuffling [78] Mn^{2+} DNase cut [79] Endonuclease V cut [80] RPR [81] RETT [82] SCRATCHY [83]
Full-length parent shuffling	Recombination of small fragments from different origin using one or more full-length parent strands; higher recombination frequency, but more elaborate	StEP [84] RACHITT [86]
Single crossover	Recombination of nonhomologous genes by ligating front and back of two different genes, selection of new genes on size; recombination possible between low or non homologous genes, but only one recombination point	(THIO)ITCHY [86,87] SHIPREC [88] SCRATCHY [83]
Domain swapping	Recombination of structural, functional, or less homologous parts of different family members; more active enzymes in the resulting library, but only a few recombination points, which are hard to find	Exon shuffling [89] DOGS [90] SISDC [91] YLBS [92] SCOPE [93]
In vivo recombination	Recombination using the gap repair system of yeast or the recE/recT system of *E. coli*; high yield, since no ligation necessary, but specialised vectors and multiple steps necessary	CLERY [94] ET-recombination [95]
Synthetic shuffling	Recombination of (un)known mutations in synthetic oligonucleotides Recombination of close mutations possible, but expensive and good selection necessary	Single-step shuffling [96] DHR [97] Synthetic shuffling [98] ADO [99]

Source: Reprinted From Ref. 73. Copyright 2005 with permission from Elsevier.

are becoming more widely used. However, developing suitable selection strategies can be extremely challenging. Darwinian evolution responds to selection pressure by seeking the easiest path. It is not obligated to follow the desired route. Thus, unanticipated and unwanted solutions may arise in the selection steps (see discussion below). If false positives become sufficiently numerous, an alternate selection process may be necessary or the system may need to be redesigned. Counter or negative selection can be applied to suppress the enrichment of unwanted catalytic activities or to fine-tune the properties of the enriched populations of catalysts.

In molecular evolution, it is critical in the design to be able to link a desired trait to a biological readout that allows identification of the encoding gene. Methods to accomplish this connection depend on the type of evolution process employed. These include directed protein evolution, phage display, and evolution of nucleic acids. As shown below, directed evolution is an extremely powerful strategy that can advance even the most feeble function into potent catalytic activity.

6.4.1. Directed Protein Evolution

The development of catalytic antibodies illustrated that it is possible to create new enzyme active sites with immunological methods. However, the chemical efficiency of these catalysts is often considerably lower than that of naturally occurring enzymes. The use of combinatorial protein chemistry is one means to overcome this deficiency. By mimicking natural evolution through recombination of genetic materials followed by selection of gene products, it is possible to engineer new enzymes. Directed evolution of proteins involves the generation of mutant gene libraries from which protein variants with valuable properties are expressed and isolated. Iterative production of improved mutants followed by mutagenesis enhances protein function through evolution, resulting in molecular breeding. Emerging proteins are studied, and clusters of mutants with optimal evolutionary potential are identified and further refined. These methods have broad applicability in protein engineering.

The starting point for in vitro evolution of enzymes has limitations imposed by the immense search space encompassed by proteins. The natural protein building blocks consist of 20 standard amino acids. Thus, a sequence of N amino acids has $20^N = {\sim}10^{1.3N}$ possible permutations. For an average protein with 300 amino acids, the entire search space contains ${\sim}10^{390}$ different sequences. From a practical aspect, this is nearly infinite. Moreover, a protein space of this magnitude will be essentially devoid of function, especially for a specific catalytic property [101]. As a corollary, beneficial mutations are rare and multiple mutations have an extremely low probability of being fruitful. Experimentally, the practical size of a completely random pool of proteins, even with relatively short sequences, will cover an infinitesimally small fraction of search space and is not likely to result in a successful search. For these reasons, most in vitro enzyme evolution approaches start with a known enzyme that is close to the desired result rather that initiating a search with a random library [101]. For example, a wildtype enzyme that catalyzes a desired transformation, but lacks the selectivity or robustness for a particular application, may serve as the parent for the first generation library of mutated proteins.

A typical approach for directed evolution of enzymes is shown in Figure 6.5 [102,103]. A mutagenesis method (Table 6.1) is used to produce the first-generation library of mutated genes from the parent gene of a wildtype enzyme. Expression vectors, produced from each of the mutant genes, are then inserted into a suitable host cell [104]. A high level of expression of mutant proteins can be difficult and unpredictable, and no universal system exists. The type of host cell, mammalian, bacterial, yeast, etc. is chosen for its ability to overexpress and often secrete the desired protein [105]. The host cells are plated onto appropriate growth media using a dilute solution so that individual cells are isolated from each other. Viable hosts contain only one expression vector and thus produce a unique mutant enzyme. The individual cells multiply into isolated colonies and consequently each colony produces one variant of the modified enzyme. In some cases, it may be possible to assay the colony directly for catalytic activity. An example of this approach was employed in the directed evolution of a *p*-nitrobenzyl esterase [106]. A substrate analog containing a *p*-nitrophenol chro-

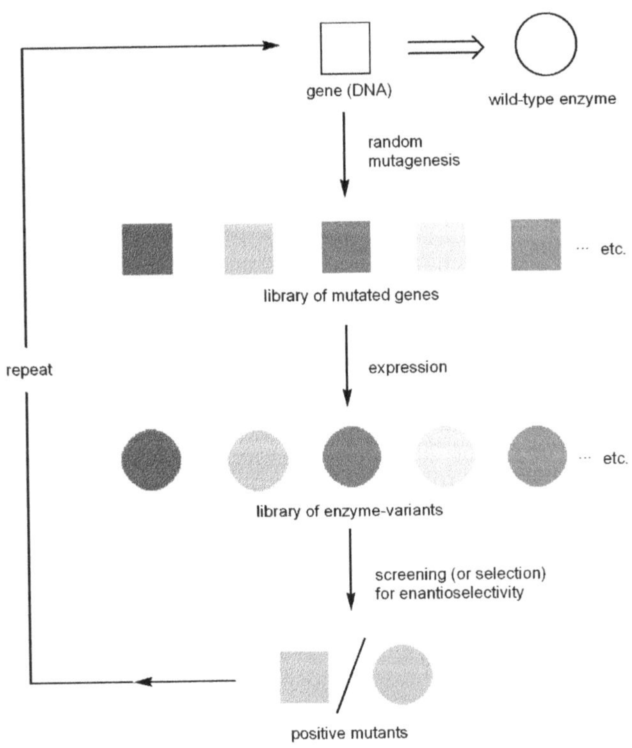

Figure 6.5. Schematic for directed evolution of an enzyme. (Reprinted by permission from Ref. 103. Copyright 2002 Wiley-VCH.)

mophore was designed that could diffuse through the cell membrane and eliminate the need for cell lysis. Cells with active enzymes hydrolyze the substrate and release the yellow-colored *p*-nitrophenol, allowing ready identification and selection for further kinetic testing. Alternatively, individual bacterial colonies can be transferred to 96- or 384-well culture plates using a robotic colony picker. The overexpressed mutant enzyme is secreted by the cell and supernatants from each well are transferred to microtiter plates where reaction screening takes place in addressable arrays. Assays based on MS, NMR, IR, and UV-visible techniques allow 1000–10,000 samples to be tested per day [107].

Once new variants are identified that have the desired characteristics, the genes for these enzymes are sequenced from their colony and subjected to additional rounds of mutagenesis, expression, and screening. In this method, the genotype and phenotype are linked by association with the cellular host. In essence, the individually separated colonies provides a spatially addressable record. Thus, a tracking method to maintain the association of a mutant enzyme with its colony is necessary. Manual processes, such as colony transfers, limit the number of mutants per day that can be assayed to ~100 per day [108]. Automation with robotics increases throughput to 1500 samples per day [102]. Prescreening methods also improve the rate at which a library can be tested. Colorometric signals such as the *p*-nitrophenol chromophore described above and other visual readout strategies provide rapid means of selecting colonies with active mutants. An additional example is illustrated by the use of glyceryl tributyrate in the agar cell growth medium in the directed evolution of lipases. Glyceryl tributyrate is a known substrate for lipases and imparts a milky appearance to the agar plates due to its insolubility. Colonies grown on glyceryl tributyrate-treated agar will reside in a clear spot on the plate if they contain active lipase mutants. In these cases, the glyceryl tributyrate is hydrolyzed by the lipase, producing water soluble products and eliminating the milky background. The number of colonies that need to be screened is considerably reduced as only the ones that produce clear spots need to be harvested.

A particularly compelling example of the power of directed evolution is its use in creating enantioselective catalysts. In developing a highly selective hydrolytic enzyme for the kinetic resolution of racemic 2-methyl decanoate, the lipase from *Pseudomonas aeruginosa* was used as a starting point (Scheme 6.5). The wildtype enzyme catalyzes the desired hydrolysis of the ester, but with low enantioselectivity (ee = 2–8%, selectivity factor $E \sim 1.1$ favoring the (*S*)-isomer). In order to monitor the library of enzyme reactions by UV-visible spectroscopy, a *p*-nitrophenol ester was used in place of the methyl version. As described previously, hydrolysis of this substrate releases the yellow-colored *p*-nitrophenylate anion (λ_{max} = 410 nm). In order to obtain chiral selectivity data, the individual (*R*)- and (*S*)-esters were used with each mutant lipase. Thus, in a rapid UV/vis screening with 96-well microtiter plates, 48 variants of the lipase could be monitored in minutes. Representative kinetic data are shown in Figure 6.6 [108]. After four rounds of evolution, a mutant enzyme with 81% ee [$E = 11.3$ for the (*S*)-ester] was identified. Exploring a

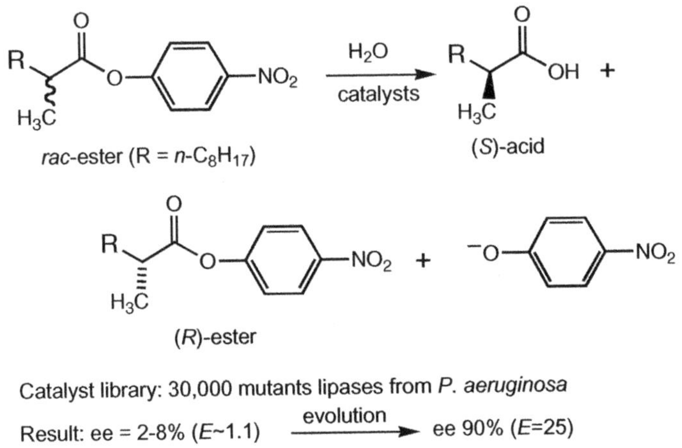

Scheme 6.5

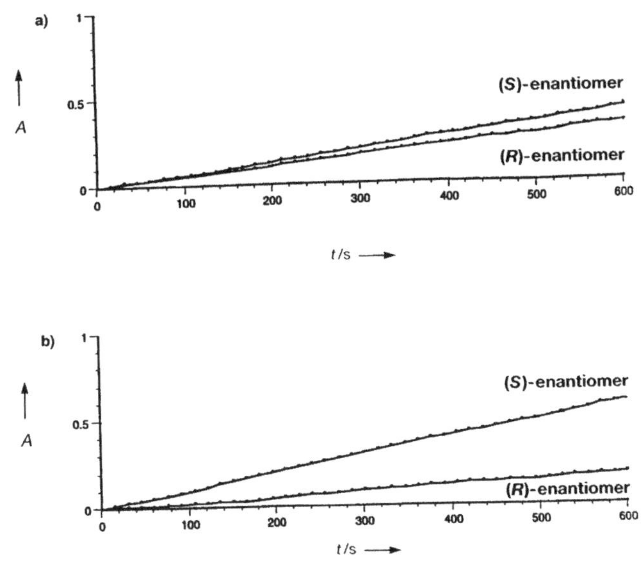

Figure 6.6. Course of lipase-catalyzed hydrolysis of (*R*)- and (*S*)-ester as a function of time: (a) wildtype from *P. aeruginosa*; (b) improved mutant from first generation. (Reprinted by permission from Ref. 108. Copyright 1997 Wiley-VCH.)

wider sequence space with improved mutatgenesis methods brought the total library size to 30,000 mutants and produced superior catalysts with ee > 90%. Moreover, the power of in vitro evolution was demonstrated by reversing the sense of the enantioselectivity and producing mutant enzymes that were selective for the *R*-ester [109,110].

A notable limitation of natural enzymes is that none are known to catalyze many of the key reactions used in commercial and academic laboratories (the

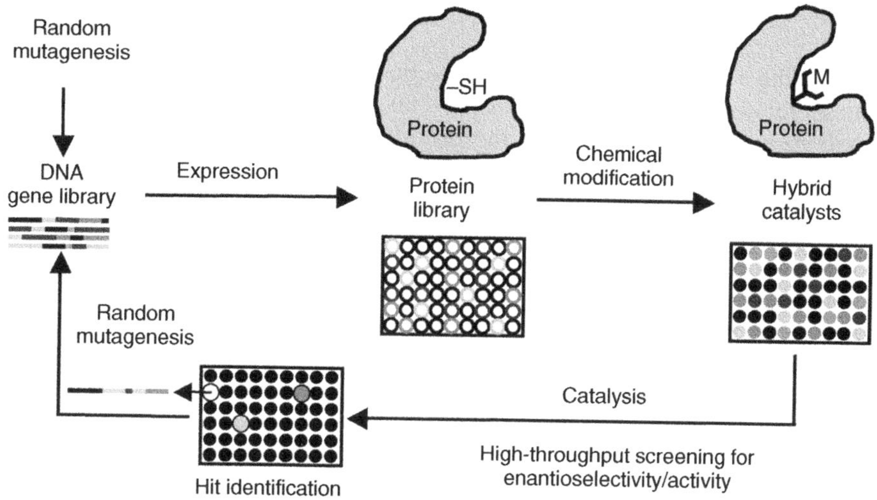

Figure 6.7. Proposed method for directed evolution of hybrid catalysts. (Reprinted by permission from Ref. 107. Copyright 2004 *Proc. Natl. Acad. Sci. USA*.)

Heck reaction, the Sonogashira reaction, hydroformylation, etc.). Thus, the application of directed evolution of proteins has no wildtype enzyme as a starting point. A potential solution to surmount this shortcoming is the concept of directed evolution of hybrid catalysts (Fig. 6.7) [107]. In this process, a pool of protein hosts would be prepared and chemically modified to bind a metal ion or metal complex to produce a library of synthetic enzymes. The gene coding for the proteins and associated molecular biology methods should allow the application of directed evolution to optimize these hybrid catalysts.

6.4.2. Phage Display

Phage display methods involve the use of biotechnology to induce the production of surface peptides fused to the coat protein of bacteriophages, viruses that infect bacteria. This is accomplished by inserting genetic information into the genomic region of the phage that codes for the coat protein(s) so that during phage morphogenesis, the encoded polypeptide is displayed on the surface of the capsid assembly of the progeny phages. The foreign peptides or proteins are expressed by infected host bacteria and must be tolerated and not be degraded in vivo. Because of its accessibility on the surface of the phage, the displayed peptide or protein can fold and behave as if it were independent of the phage. Once the phage infects a cell, it produces progeny by parasitically employing the host's cellular processes. In nonlytic cases, progeny phages are secreted continuously from the infected cell without killing the host. Each division cycle produces several hundred phages per host cell.

Cloning a library of genes produces a pool of chimeric phages that contain the targeted library of peptides or proteins. In the phage library, each phage contains a unique foreign DNA insert and thus displays a unique peptide on its surface. The power of this approach derives from the physical linking of genetic information and its expression product through a phage particle. This allows the gene of the selected protein to be sequenced, cloned, and expressed. Since the development of phage display in 1985, this method has been used extensively in the discovery of peptides that bind receptors, enzyme inhibitors, and protein binding sites for ligands such as high affinity antibodies [111]. On infecting a fresh bacterial host cell, a phage is able to produce a large number of identical copies of itself. Moreover, mutation of the extra genetic sequence results in the display of a mutated peptide. These characteristics of replication and mutation provide the means for in vitro evolution.

The most commonly used phage display systems are based on filamentous bacteriophages (Fig. 6.8) f_1, f_d, M_{13}, and related phagemids [112,113]. These viruses contain single-stranded DNA and infect *E. coli.* In addition, they tolerate a wide pH range of 2.5–11 and temperatures up to 80°C. Phage display methods have also been implemented with lambda and T4 phages [114]. Proteins as large as 86 kDa have been expressed on f_d with a loading of one copy per phage [115]. Small peptides can be expressed in up to three to four copies per coat protein, producing several thousands of foreign peptide per phage particle. Three to five copies of pIII are located at the tip of the phage and about 2500 copies of pVIII make up the coat protein. The phage enzymes can be purified by precipitation with polyethylene glycol followed by elimination of the bacteria.

Development of catalysts using phage display can be accomplished by a variety of strategies, including the same method as that used for catalytic antibodies, but with important differences. Although antibodies are high affinity binders for a transition state analog, the best binders may not function as an optimal catalyst. Phage display technology allows for the use of evolution-

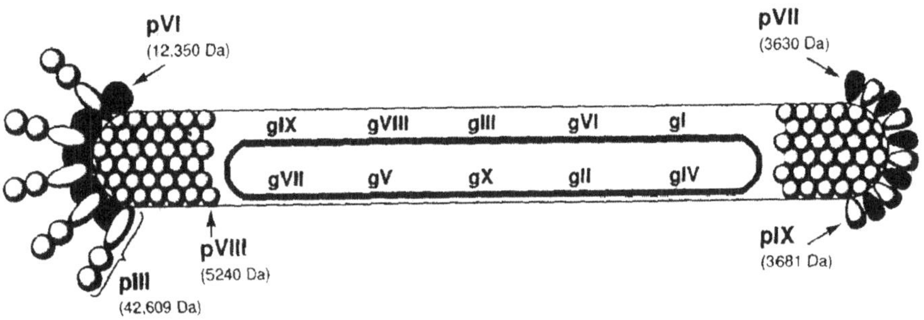

Figure 6.8. Schematic of a filamentous phage particle. (Reprinted by permission from Ref. 69. Copyright 2005 Wiley-VCH.)

ary pressure to select for catalytic function and not solely for strong binding. Figure 6.9 illustrates the use of affinity screening of a phage library. By increasing a selection criterion, such as lower temperature or time, displayed peptides can be isolated that have higher binding for the TSA and can be subsequently amplified. Mutations are introduced periodically to increase the diversity of the protein library. When a suitable binding level is achieved, individual phages are cloned and the displayed peptide is tested for its catalytic efficiency. The advantage of phage display over catalytic antibodies is that animal immunization is not required and is, therefore, faster than hybridoma techniques [116]. A potential disadvantage of phage display approach is that a high level of display is not always tolerated by the host.

Phage display methods can be employed with catalytic antibodies. This was demonstrated using antibody 17E8 for the hydrolysis of phenyl esters to

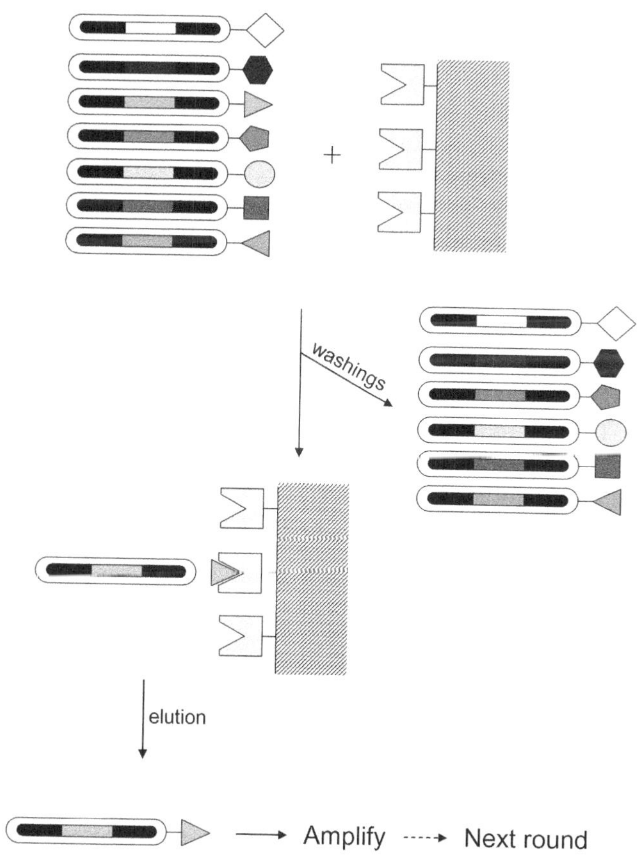

Figure 6.9. Phage display approach based on in vitro selection of proteins expressed on the surface of filamentous phages. (Reprinted by permission from Ref. 117. Copyright 2005 Wiley-VCH.)

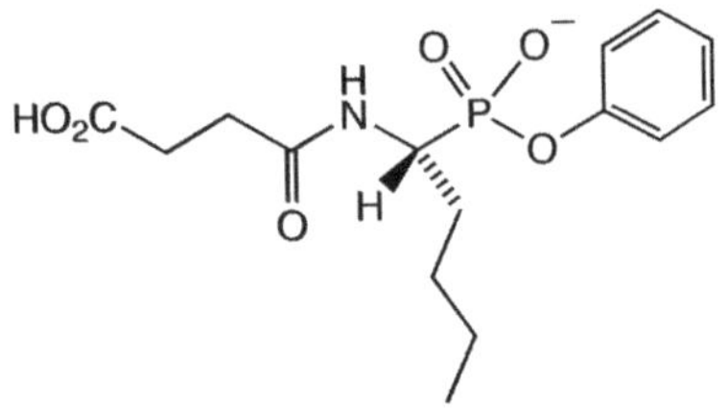

Figure 6.10. Phosphonate ester used for eliciting catalytic antibody.

carboxylic acids [Eq. (6.5)] [118]. The phosphonate transition state analog used in the eliciting the displayed antibody is shown in Figure 6.10.

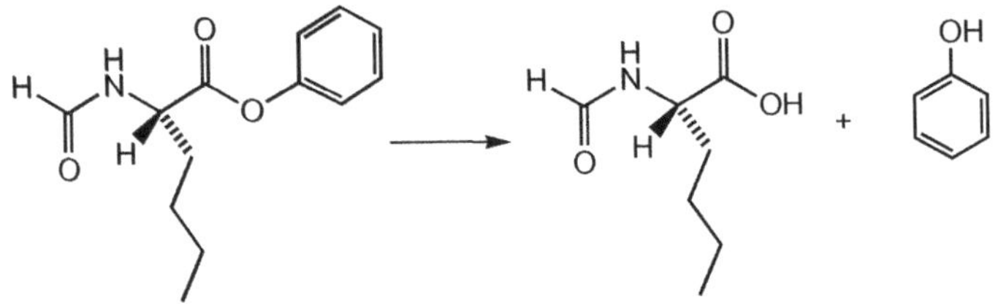

(6.5)

Suicide inhibitors provide a means for selecting for catalytic activity. In these cases, reaction within the enzyme active site converts the substrate into a reactive intermediate that rapidly forms a covalent bond with a proximal amino acid. This irreversibly inhibits the enzyme and allows the immobilization of the active mutants if the inhibitor is linked to a solid support. This approach was used in the optimization of a phosphate hydrolysis enzyme from a highly diverse HuCAL-scFv library [119]. Figure 6.11a shows the suicide inhibitor, and Figure 6.11b illustrates the mechanism of the suicide inhibition followed by selection. This approach uses direct turnover-based selection from an unbiased library. The result obtained in this manner was 10^2-fold better than that of the best catalytic antibody elicited for phosphate ester hydrolysis.

6.4.3. Nucleic Acids

The ability of RNA to function as an enzyme was discovered by Cech [120,121] and Altman [122]. Since nucleic acids possess both properties of information content (genotype) and function (phenotype) in a single molecule, RNA and its deoxy form, DNA, are ideal candidates for in vitro evolution [123]. The genotype is stored in its sequence and is recognized by polymerases, enabling these enzymes to produce copies of the original sequence with high fidelity. The sequence also dictates the 3D structure that governs functional (phenotypic) properties, including catalysis. A key attribute is that the deconvolution strategies needed for large libraries is an inherent property of nucleic acids. Thus, in vitro evolution is greatly simplified with nucleic acids as functional genetic material can be mutated, selected, and amplified without the need to

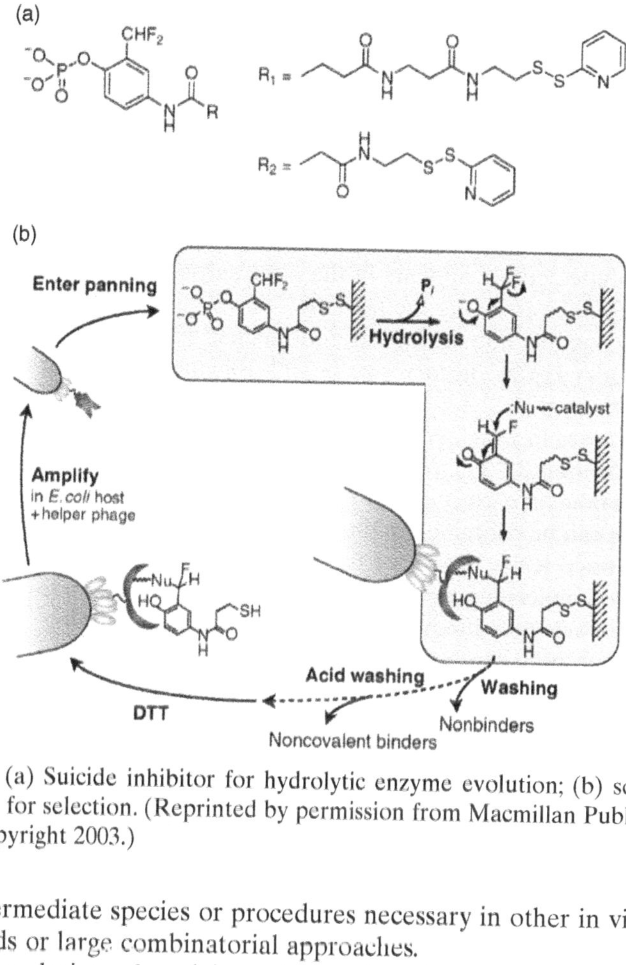

Figure 6.11. (a) Suicide inhibitor for hydrolytic enzyme evolution; (b) schematic of process used for selection. (Reprinted by permission from Macmillan Publishers Ltd.: Ref. 119. Copyright 2003.)

involve intermediate species or procedures necessary in other in vitro evolution methods or large combinatorial approaches.

In vitro evolution of nucleic acids was first developed to produce small RNA molecules for desired functions such as selective binding of molecules and proteins [124–126]. The general approach involves a process commonly known as "*s*ystematic *e*volution of *l*igands by *ex*ponential enrichment" (SELEX) [126]. This protocol is a promising tool for developing nucleic acids for a wide range of uses such as new diagnostic and therapeutic reagents. With SELEX, large random pools of nucleic acids can be screened for a desired functionality. The term *aptamers*, used for these functional nucleic acids, is derived from the Latin word *aptus* (meaning to fit) and the Greek word *meros* (part). The use of SELEX in the evolution of nucleic acid-based catalysts, will be discussed here.

A useful feature in the application of nucleic acids is the ability to program automated synthesizers to randomly incorporate nucleotides in specific sections in the molecule. Thus, large libraries can be chemically fabricated to contain DNA strands with completely random sequences incorporated. The manipulated strands also have conserved regions at each end that serve as recognition sites for polymerases. The synthesized library of DNA is amplified a few times by the polymerase chain reaction (PCR) [127] to produce several copies of each strand. For DNA evolution, this amplified library serves as the initial pool. For RNA studies, the library is transcribed from the random DNA to generate the RNA pool. This produces a library with immense diversity, allowing a combinatorial starting point that makes no assumptions about the nucleotide composition needed to achieve a particular function. Once the desired catalytic activity appears in the pool, subsequent rounds of evolution can be undertaken to refine the activity.

The total possible number of different nucleic acid molecules for an N-mer is 4^N ($10^{0.6N}$). A typical library size is 10^{14}–10^{15} molecules. For $N = 25$, total possible variations is 1.1×10^{15}. The entire possible search space can be covered for cases in which $N < 25$. However, catalytic function may require N to be greater than 25. For $N = 85$, more than 10,000 earth masses would be needed to make one copy of all permutations. Thus only a portion of the search space can be contained in the typical library. In vitro selection has been used to identify RNA aptamers for binding a variety of targets. However, catalytic RNA molecules are generally longer sequences than simple aptamers developed for selective binding. Thus, the probability of finding catalytic activity in a random pool can be extremely small. However, once activity is found, the use of mutagenic methods (Table 6.1) allows the diversity of the library to be increased such that additional sequence space can be covered that is widely disparate from the initial pool.

The few nucleic acid enzymes that occur in nature are all RNA enzymes (ribozymes) that generally catalyze phosphate chemistry. These reactions typically involve RNA cleavage, RNA ligation (extension), phosphotransfer, phosphoester hydrolysis, etc. The exception is peptidyl transfer catalyzed by the ribozyme of the ribosomal subunit. Not surprisingly, a majority of the artificially evolved ribozymes has been selected for phosphoester transfer reactions. An example of a ribozyme selection for a ligase reaction is illustrated in Figure 6.12 [128]. In order to find the rare active species in large libraries of mostly inactive materials, a critically designed selection step is needed to identify the desired trait among a large number of nonfunctional species. A typical procedure is to utilize a recognition tag. This tag may be a unique oligonucleotide sequence that can be captured by hybridization to its nucleic acid complement or an affinity label such as biotin that can be retained by strong binding to streptavidin immobilized on an affinity column [129]. Catalytically active nucleic acids are selected from the pool as a result of self-modification involving the recognition tag resulting from the desired reaction. In a bond forming reaction, a substrate is attached to the nucleic acid and the reactant is linked to the tag. Thus, on successfully promoting the desired reaction, the active nucleic acid strands undergo self-tagging. The catalytic species would then be recovered through specific recognition of the tag. In contrast, for a cleavage reaction, the tagged substrate is chemically linked to the nucleic acids. On bond breaking, the active

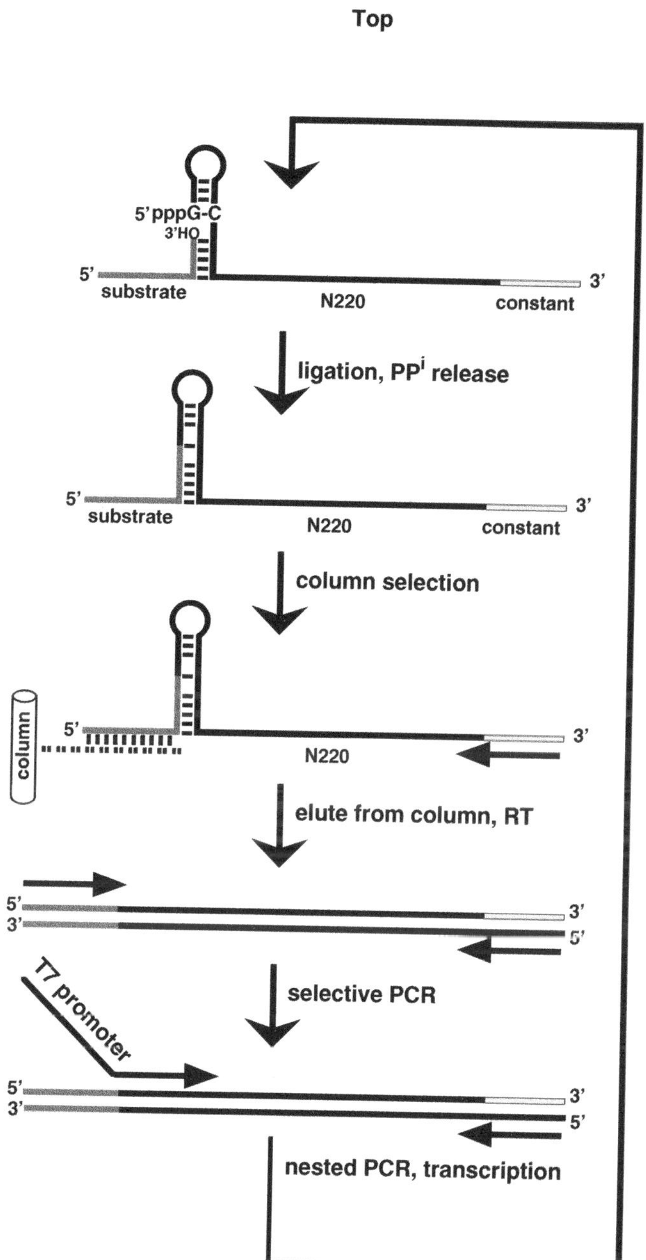

Figure 6.12. Schematic for RNA ligase ribozyme selection. (Reprinted, with permission, from the *Annual Review of Biochemistry*, Vol. 68 © 1999 by *Annual Reviews*, www.annualreviews.com).

species lose their tag and inactive sequences are removed through a selection process that is tag-specific. Although technically these processes are single-turnover and not truly catalytic, it is generally possible to convert the self-modifying nucleic acid into a true catalyst by removing the substrate [121].

Selection methods used with nucleic acids are extremely powerful methods for identifying desired traits in large libraries. Unlike protein enzymes, where the search space is so immense that it is better to start with a native enzyme and use directed evolution to develop new catalytic function, with nucleic acid enzymes it is possible to start from completely random sequences to create new catalysts. Most random RNA sequences are often highly structured because of the high probability of Watson–Crick base pairing forming in a random string [130]. This statistical likelihood results in a relatively large collection of 3D shapes in a random RNA library. Given the large number of different shapes, the assumption is made that a random sample size of 10^{15} will contain some species with catalytic activity [131,132]. In addition to eliciting catalytic activity from a completely random library, particular characteristics of the catalysis can be refined, such as rate, substrate selectivity, and tolerance for particular reaction conditions. Thus, to develop catalysts capable of high rates, a selection pressure of decreasing reaction times for each evolutionary cycle is employed. Although selection conditions may be directed toward a particular trait, unintended outcomes are always a possibility as nature is extremely clever at deriving a solution. For example, in selection for a cleavage catalyst that is identified by loss of a recognition tag, some nucleic acids may be falsely isolated because they have inadvertently masked the tag rather than induced the detachment. Alternatively, loss of the tag may occur due to cleavage of a phosphate backbone site upstream of the intended substrate or target site [3]. In these cases, counter selection methods must be devised to eliminate the undesirable traits if the evolutionary process is to be continued.

The potential for unintended selection is illustrated by attempts to develop a histidine-dependent DNA enzyme for cleaving RNA. When 20 mM L-histidine and 0.5 mM $MgCl_2$ were used to trigger the reaction, the evolved catalysts employed Mg^{2+} instead of histidine as the catalytic cofactor [133]. The large concentration advantage of histidine over Mg^{2+} is insufficient to overcome problems in positioning histidine correctly to assist in the cleavage. Exclusion of divalent metals from the selection process was necessary in order to evolve histidine-dependent catalysts [134].

One initial application of in vitro evolution of nucleic acids was developing aptamers for selective binding of specific targets. This process is analogous to the antibody immune response and suggested that the approach used for catalytic antibodies should work for catalytic aptamers. However, the in vitro selection of aptamers to bind a transition state analogue generally has not led to successful development of new catalytic species [131]. This may be due to a lack of good transition state analogs designed to bind with nucleic acids. Among the cases that have succeeded are selection of aptamers for binding targets that mimic a geometric distortion needed to reach a transition state. Recent examples involve nucleic acid enzymes that catalyze the insertion of Cu^{2+} and Zn^{2+}

into porphyrins [135,136]. This approach is similar to the development of catalytic antibodies for the same porphyrin metallation reaction (see discussion above). In both cases, nucleic acids with similar catalytic efficiencies were evolved. A 35-nucleotide RNA enzyme, RNA + 12.19, exhibited characteristic enzyme (Michaelis–Menten) kinetics [137] with a specificity constant $k_{cat}/K_M = 1100\,M^{-1}s^{-1}$ [135]. In comparison, the catalytic antibody specificity constant for the same porphyrin metallation was $450\,M^{-1}s^{-1}$ (Section 6.3).

Unlike RNA enzymes (ribozymes), no catalytic DNA species have been found in nature [138]. This is likely due to the existence of DNA primarily in its well-known double helical form. The binding of two complementary DNA strands through Watson–Crick base pairing [139] would restrict catalytic properties much as an antisense oligonucleotide inactivates a ribozyme [140]. In isolated single-stranded form, DNA is able to fold into intricate 3D structures that provide useful functional properties, much like RNA aptamers. However, DNA aptamers and RNA aptamers based the same sequences do not function equivalently [131]. Such intrinsic differences may arise from hydrogen-bonding factors related to the presence or absence of the 2′-OH group on the ribose (Fig. 6.13). For example, DNA is less likely to form structures such as "ribose zippers." [141] Nonetheless, molecular evolution approaches have created DNA enzymes (deoxyribozymes), illustrating that the lack of the 2′-OH is not a serious handicap. The range of DNA enzymes lags behind that of RNA, but this simply illustrates largely untapped potential. At this point, any functional handicap of DNAzymes do not seem to be due to limitations in the nature of their active sites, but by the methods used to obtain them [140]. Moreover, DNA has some key practical advantages over RNA [138]: (1) DNA is approximately sevenfold less expensive that RNA, (2) sequence lengths for DNA can generally be made longer than for RNA, (3) procedures needed to manipulate

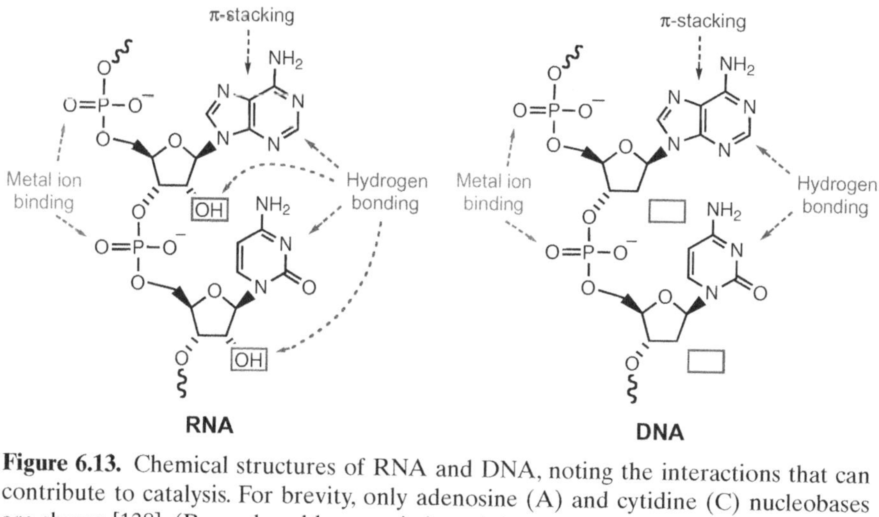

Figure 6.13. Chemical structures of RNA and DNA, noting the interactions that can contribute to catalysis. For brevity, only adenosine (A) and cytidine (C) nucleobases are shown [138]. (Reproduced by permission of the Royal Society of Chemistry.)

DNA are more efficient than for RNA as transcription and reverse transcription steps are not needed, and (4) DNA has a chemical stability greater than that of RNA. Under physiological conditions, the half-life for phosphate backbone hydrolysis of DNA is greater than 10^6 years [142]. This is five orders of magnitude longer than for the same cleavage of RNA.

An example of a metal-dependent nucleic acid catalyst is the Co^{2+}-dependent DNA enzyme for cleaving RNA. In order to improve specificity for Co^{2+} as the transition metal cofactor, a negative selection protocol was included to avoid unwanted competing metal ions. Figure 6.14 illustrates the utility of the negative selection strategy. Without the negative selection steps, evolved catalytic DNA species were more active with Zn^{2+} and Pb^{2+} instead of with Co^{2+}. However, including both positive (selecting for Co^{2+}) and negative (eliminating competing metal ions) selection produced catalytic DNA that was more active with Co^{2+} [143].

A dramatic example of the power of molecular evolution is the production of allosteric ribozymes. In these catalysts, ligand binding at one site controls the catalytic activity of a distant site [144]. Figure 6.15a depicts a Co^{2+}-dependent self-cleaving allosteric ribozyme (AR1). No phosphate ester cleavage in the AR1 backbone occurs until Co^{2+} is bound in the allosteric domain (Fig. 6.15d) [145].

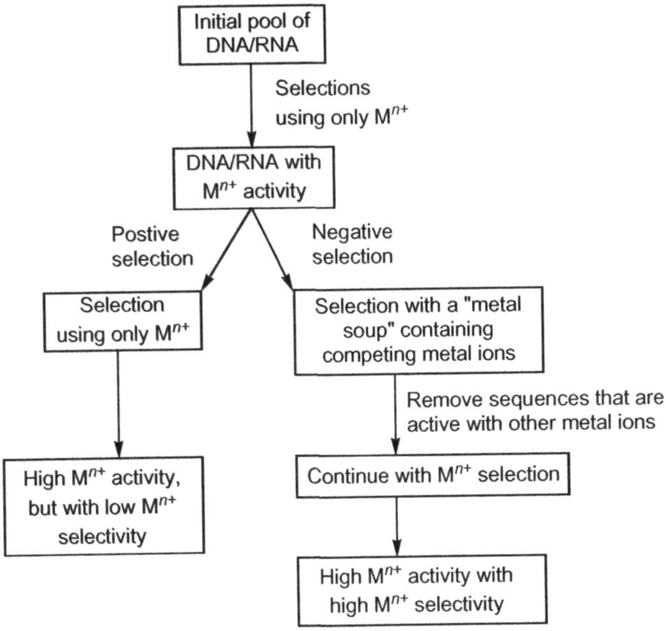

Figure 6.14. Evolution strategies for improving metal ion selectivity with positive and negative selection steps. (Reprinted by permission from Ref. 143. Copyright 2002 Wiley-VCH.)

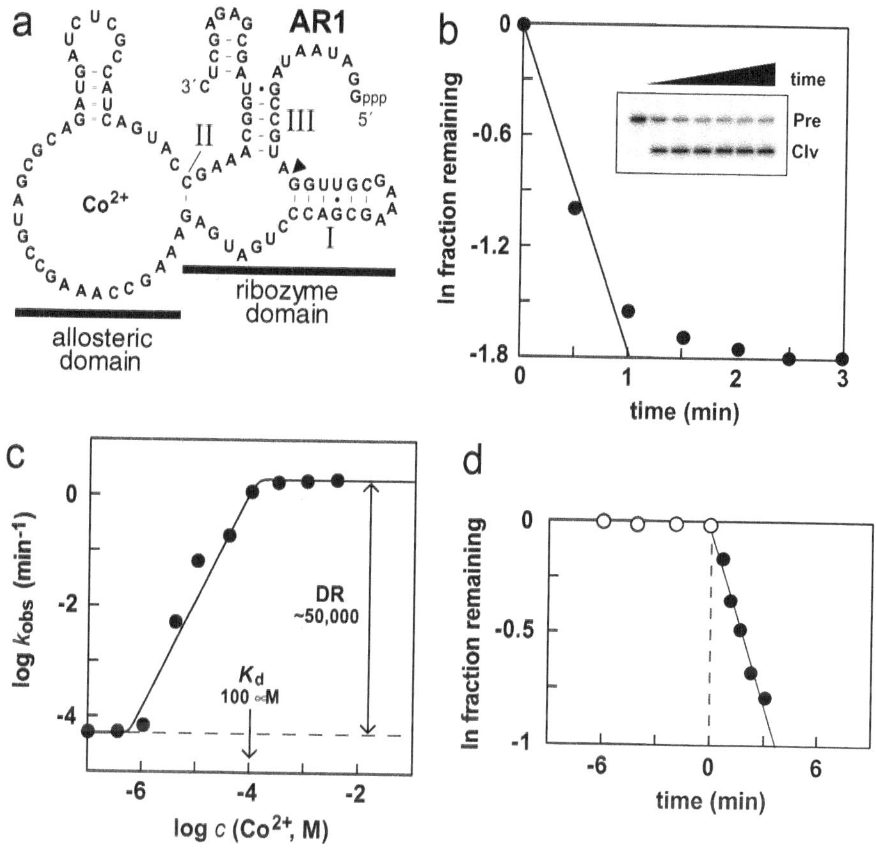

Figure 6.15. (a) Secondary-structure model for a cobalt-sensitive allosteric ribozyme (AR1) with distinct allosteric and ribozyme domains indicated. The arrowhead identifies the site of self-cleavage. (b) Time-dependent cleavage of 5′-^{32}P-labeled AR1 (trace) in the presence of 500 μM $CoCl_2$ under standard conditions. (c) Kinetic properties of AR1 with various concentrations of Co^{2+}. The dashed line reflects the k_{obs} for ribozyme cleavage in the absence of added Co^{2+} (DR = dynamic range). (d) Rapid activation of the AR1 ribozyme on the addition of 1 mM $CoCl_2$ (dashed line) at 50°C under standard conditions. Open and closed circles represent data collected in the absence and presence of $CoCl_2$, respectively. (Reprinted by permission from Macmillan Publishers Ltd.: Ref. 145. Copyright 2001.)

Recently, the use of DNA-based enzymes was developed for catalyzing the Heck and Sonagashira reactions [Eqs. (6.6) and (6.7), respectively] [146]. From a library of 10^{14} different variants of single-stranded DNA with a random section of 40 nucleotides, catalysts that effectively promote these C—C bond-forming reactions were evolved [147]. Several rounds of selection, including one mutagenesis step, produced 16 sequences for each reaction that increased the catalytic efficiency by more than 10^5 over that of the initial library. The

Heck and Sonagashira reactions are not natural biological processes and thus not catalyzed by enzymes. These DNAzymes illusrate the potential for practical applications and use in contemporary organic synthesis techniques [148].

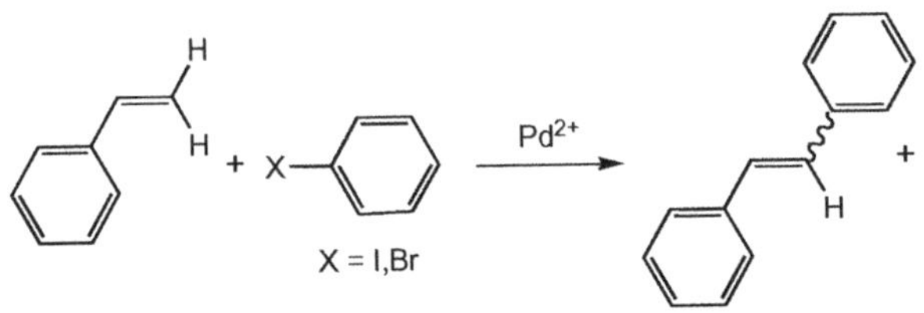

(6.6)

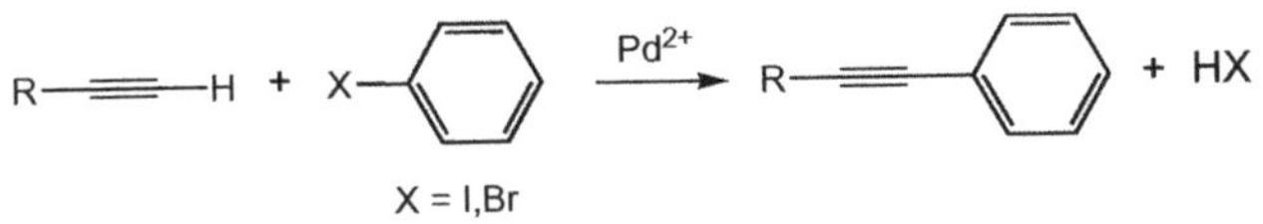

(6.7)

The small number (four) of subunits for nucleic acids places a significant restriction on the functionality of RNA and DNA relative to amino acid-based structures that have twenty building blocks available. Notably, nucleic acids lack a general acid base with a pK_a that is near neutral (as in histidine), a primary alkyl amine (as in lysine), a carboxylate (as in aspartate), and a sulfhydryl (as in cysteine) [149]. However, chemists are not constrained by natural design and new functionality can be imparted by employing modified nucleotides or adding a small-molecule cofactor. A number of unnatural nucleotides have been developed (Fig. 6.16) that can create new transition-metal-binding sites in nucleic acid sequences. Some of these nucleotides are tolerated by PCR enzymes and can be manipulated by typical molecular biological methods [150–152]. With these new building blocks, the directed evolution approach will be expanded to a wider range of chemical possibilities and potential for developing and understanding new catalytic systems beyond what nature can provide.

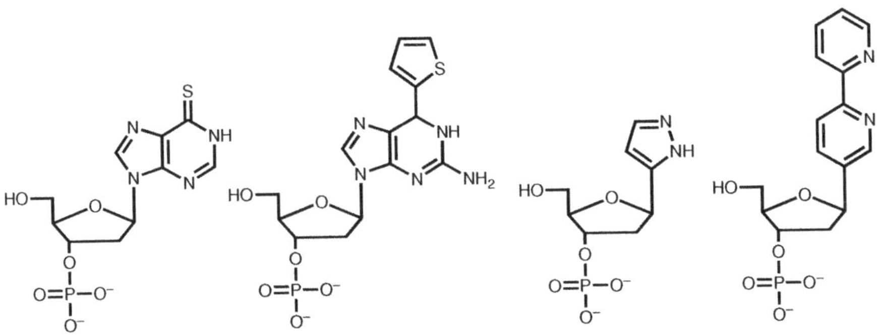

Figure 6.16. Examples of unnatural nucleotides with metal binding sites that have been developed for expanding the functionality of nucleic acids.

6.5. GENERAL PERSPECTIVES

The application of combinatorial techniques and molecular evolution approaches in the field of homogeneous catalysis is still in its infancy. However, powerful approaches are emerging and increasing in scope. When fully harnessed, these will allow scientists the ability to manipulate and understand chemical processes at their most fundamental level. Further developments, such as improved miniaturization, automation, posttranscriptional modification, and continuous in vitro evolution [153] to minimize user intervention will accelerate the widespread application and acceptance of these strategies. Moreover, the capabilities of rational design, particularly computational techniques and de novo design, are expanding also. The use of all of these methods will be instrumental in the development and study of catalysts. It is clear that the true potential for sophisticated processes is far greater than presently exhibited and it is likely that many novel and unanticipated catalysts will be created and identified by combinatorial techniques and directed evolution. These methods will not only yield a greater understanding of catalytic performance and practical means of achieving important chemical transformations but will also ultimately lead to a better understanding of complex chemical processes. Although this review only provides a snapshot of a rapidly growing field and many aspects of are not covered, it is hoped that sufficient material has been introduced to give scientists enough background to begin thinking about applying combinatorial approaches or directed evolution methods to new systems.

REFERENCES

1. Archibald, B., Brummer, O., Devenney, M., Gorer, S., Jandeleit, B., Uno, T., Weinberg, W. H., and Weskamp, T., Combinatorial methods in catalysis, in *Handbook of Combinatorial Chemistry*, K. C. Nicolaou, R. Hanko, and W. Hartwig (eds.), Wiley-VCH, Weinheim, 2002, Vol. 2, pp. 885–990.
2. Brakmann, S. and Johnsson, K., *Directed Molecular Evolution of Proteins*, Wiley-VCH, Weinheim, 2002.
3. Joyce, G. F., Directed evolution of nucleic acid enzymes, *Annu. Rev. Biochem.* **73**:791–836 (2004).
4. Keinan, E., *Catalytic Antibodies*, Wiley-VCH, Weinheim, 2005.
5. Reetz, M. T., Combinatorial methods in catalysis by metal complexes, in *Comprehensive Coordination Chemistry II*, J. A. McCleverty and T. J. Meyer (eds.), Elsevier, Amsterdam, 2004, Vol. 9, pp. 509–548.
6. Svendsen, A., *Enzyme Functionality: Design, Engineering, and Screening*, Marcel Dekker, New York, 2004.
7. Achenbach, J. C., Chiuman, W., Cruz, R. P. G., and Li, Y., DNAzymes: From creation in vitro to application in vivo, *Curr. Pharm. Biotechnol.* **5**(4):321–336 (2004).

8. Trost, B. M., The atom economy—a search for synthetic efficiency, *Science* **254**:1471–1477 (1991).
9. Hagemeyer, A., Strasser, P., Volpe, J., and Anthony F., *High-Throughput Screening in Chemical Catalysis*, Wiley-VCH, Weinheim, 2004.
10. Parshall, G. W. and Ittel, S. D., *Homogeneous Catalysis*, Wiley-Interscience, New York, 1992.
11. Brown, J. M., Nobel Prizes in chemistry; asymmetric hydrogenation recognised! *Adv. Synth. Catal.* **343**(8):755–756 (2001).
12. Marko, I. E., Nobel Prizes in chemistry 2001; asymmetric oxidation rewarded! *Adv. Synth. Catal.* **343**(8):757–758 (2001).
13. Nobel Prizes 2005: Chemistry, physiology or medicine, physics, *Angew. Chem. Int. Ed* **44**(43):6982 (2005).
14. van Leeuwen, P. W. N. M., *Homogeneous Catalysis: Understanding the Art*, Kluwer, Dordrecht, 2004.
15. Knowles, W. S., Asymmetric hydrogenations—the Monsanto L-Dopa process, in *Asymmetric Catalysis on Industrial Scale*, H.-U. Blaser and E. Schmidt (eds.), Wiley-VCH, Weinheim, 2004, pp. 23–38.
16. Hahn, K. W., Klis, W. A., and Stewart, J. M., Design and synthesis of a peptide having chymotrypsin-like esterase activity, *Science* **248**(4962):1544–1547 (1990).
17. Breslow, R., *Artificial Enzymes*, Wiley-VCH, Weinheim, 2005.
18. Murakami, Y., Kikuchi, J.-I., Hisaeda, Y., and Hayashida, O., Artificial enzymes, *Chem. Rev.* **96**:721–758 (1996).
19. Wang, Y., DuBois, J. L., Hedman, B., Hodgson, K. O., and Stack, T. D. P., Catalytic galactose oxidase models: Biomimetic Cu(II)-phenoxyl-radical reactivity, *Science* **279**(5350):537–540 (1998).
20. Chaudhuri, P., Hess, M., Florke, U., and Wieghardt, K., From structural models of galactose oxidase to homogeneous catalysis: Efficient aerobic oxidation of alcohols, *Angew. Chem. Int. Ed.* **37**(16):2217–2220 (1998).
21. Schmid, A., Dordick, J. S., Hauer, B., Kiener, A., Wubbolts, M., and Witholt, B., Industrial biocatalysis today and tomorrow, *Nature* **409**:258–268 (2001).
22. Zaks, A., Industrial catalysis, *Curr. Opin. Chem. Biol.* **5**:130–136 (2001).
23. Liese, A. and Filho, M. V., Production of fine chemicals using biocatalysis, *Curr. Opin. Biotechnol.* **10**:595–603 (1990).
24. Gerlt, J. A., Relationships between enzymatic catalysis and active site structure revealed by applications of site-directed mutagenesis, *Chem. Rev.* **87**:1079–1105 (1987).
25. Knowles, J. R., Tinkering with enzymes: What are we learning, *Science* **236**(4806):1252–1258 (1987).
26. Benkovic, S. J., Fierke, C. A., and Naylor, A. M., Insights into enzyme function from studies on mutants of dihydrofolate reductase, *Science* **239**(4844):1105–1110 (1988).
27. Hirose, Y., Kariya, K., Nakanishi, Y., Kurono, Y., and Achiwa, K., Inversion of enantioselectivity in hydrolysis of 1,4-dihydropyridines by point mutation of lipase PS, *Tetrahedron Lett.* **36**:1063–1066 (1995).

28. Lehmann, M., Concepts for protein engineering, in *Enzyme Functionality: Design, Engineering, and Screening*, A. Svendsen (ed.), Marcel Dekker, New York, 2004, pp. 1–14.

29. Ongeri, S., Piarulli, U., Jackson, R. F. W., and Gennari, C., Optimization of new chiral ligands for the copper-catalysed enantioselective conjugate addition of Et_2Zn to nitroolefins by high-throughput screening of a parallel library, *Eur. J. Org. Chem.* **2001**(4):803–807 (2001).

30. Zhang, Y., Gong, X., Zhang, H., Larock, R. C., and Yeung, E. S., Combinatorial screening of homogeneous catalysis and reaction optimization based on multiplexed capillary electrophoresis, *J. Combin. Chem.* **2**:450–452 (2000).

31. Long, J., Hu, J., Shen, X., Ji, B., and Ding, K., Discovery of exceptionally efficient catalysts for solvent-free enantioselective hetero-Diels-Alder reaction, *J. Am. Chem. Soc.* **124**(1):10–11 (2002).

32. Geysen, H. M., Meloen, R. H., and Barteling, S. J., Use of peptide synthesis to probe viral antigens for epitopes to a resolution of a single amino acid, *Proc. Natl. Acad. Sci. USA* **81**(13):3998–4002 (1984).

33. Houghten, R. A., General method for the rapid solid-phase synthesis of large numbers of peptides: Specificity of antigen–antibody interaction at the level of individual amino acids, *Proc. Natl. Acad. Sci. USA* **82**(15):5131–5135 (1985).

34. Erb, E., Janda, K. D., and Brenner, S., Recursive deconvolution of combinatorial chemical libraries, *Proc. Natl. Acad. Sci. USA* **91**:11422–11426 (1994).

35. Coffen, D. L. and Luithle, J. E., Introduction to combinatorial chemistry, in *Handbook of Combinatorial Chemistry*, K. C. Nicolaou, R. Hanko, and W. Hartwig (eds.), Wiley-VCH, Weinheim, 2002, Vol. 1, pp. 10–23.

36. Liu, G. and Ellman, J. A., A general solid-phase synthesis strategy for the preparation of 2-pyrrolidinemethanol ligands, *J. Org. Chem.* **60**(24):7712–7713 (1995).

37. Sigman, M. S. and Jacobsen, E. N., Schiff base catalysts for the asymmetric Strecker reaction identified and optimized from parallel synthetic libraries, *J. Am. Chem. Soc.* **120**(19):4901–4902 (1998).

38. Andres, C. J., Whitehouse, D. L., and Deshpande, M. S., Transition-metal-mediated reactions in combinatorial synthesis, *Curr. Opin. Chem. Biol.* **2**(3):353–362 (1998).

39. Gilbertson, S. R., Combinatorial-parallel approaches to catalyst discovery and development, *Progress Inorg. Chem.* **50**:433–471 (2001).

40. Feher, M. and Schmidt, J. M., Property distributions: Differences between drugs, natural products, and molecules from combinatorial chemistry, *J. Chem. Inform. Comput. Sci.* **43**(1):218–227 (2003).

41. Pauling, L., Molecular architecture and biological reactions, *Chem. Eng. News* **24**:1375–1377 (1946).

42. Jencks, W. P., *Catalysis in Chemistry and Enzymology*, McGraw-Hill, New York, 1969, p. 288.

43. Pollack, S. J., Jacobs, J. W., and Schultz, P. G., Selective chemical catalysis by an antibody, *Science* **234**(4783):1570–1573 (1986).

44. Tramontano, A., Janda, K. D., and Lerner, R. A., Catalytic antibodies, *Science* **234**:1556–1570 (1986).
45. Wilson, I. A. and Stanfield, R. L., Antibody-antigen interactions, *Curr. Opin. Struct. Biol.* **3**:113–118 (1993).
46. Davies, D. R., Padlan, E. A., and Sheriff, S., Antibody-antigen complexes, *Annu. Rev. Biochem.* **59**:439–473 (1990).
47. Dailey, H. A., Dailey, T. A., Wu, C. K., Medlock, A. E., Wang, K. F., Rose, J. P., and Wang, B. C., Ferrochelatase at the millennium: Structures, mechanisms and [2Fe-2S] clusters, *Cell Mol. Life Sci.* **57**(13–14):1909–1926 (2000).
48. Lavallee, D. K. and Anderson, O. P., Crystal and molecular structure of a free-base N-methylporphyrin; N-methyl-5,10,15,20-tetra(p-bromophenyl)porphyrin, *J. Am. Chem. Soc.* **104**(17):4707–4708 (1982).
49. Cochran, A. G. and Schultz, P. G., Antibody-catalyzed porphyrin metallation, *Science* **249**:781–783 (1990).
50. Rajewsky, K., Forester, I., and Cumano, A., Evolutionary and somatic selection of the antibody repertoire in the mouse, *Science* **238**(4830):1088–1094 (1987).
51. Alt, F. W., Blackwell, T. K., and Yancopoulos, G. D., Development of the primary antibody repertoire, *Science* **238**(4830):1079–1087 (1987).
52. Honjo, T. and Habu, S., Origin of immune diversity: Genetic variation and selection, *Annu. Rev. Biochem.* **54**:803–830 (1985).
53. Burton, D. R., Overview: Amplification of antibody genes, in *Phage Display: A Laboratory Manual*, C. F. Barbas III, D. R. Burton, J. K. Scott, and G. J. Silverman (eds.), Cold Spring Harbor Laboratory Press, New York, 2001, pp. 8.1–8.3.
54. Köhler, G. and Milstein, C., Continuous cultures of fused cells secreting antibody of predefined specificity, *Nature* **256**(5517):495–497 (1975).
55. Thomas, N. R., Hapten design for the generation of catalytic antibodies, *Appl. Biochem. Biotechnol.* **47**:345–372 (1994).
56. Stewart, J. D. and Benkovic, S. J., Transition-state stabilization as a measure of the efficiency of antibody catalysis, *Nature* **375**(6530):388–391 (1995).
57. Nakayam, G. R. and Schultz, P. G., Approaches to the design of semisynthetic metal-dependent catalytic antibodies, in *Catalytic Antibodies*, D. J. Chadwick and J. Marsh (eds.), Wiley, Chichester, UK, 1991; Ciba Foundation Symp. 159, pp. 72–90.
58. Janda, K. D., Weinhouse, M. I., Danon, T., Pacelli, K. A., and Schloeder, D. M., Antibody bait and switch catalysis: A survey of antigens capable of inducing abzymes with acyl-transfer properties, *J. Am. Chem. Soc.* **113**:5427–5434 (1991).
59. Wilson, I. A., Stanfield, R. L., Rini, J. M., Areval, J. H., Schulze-Gahmen, U., Fremont, D. H., and Stura, E. A., Structural aspects of antibodies and antibody-antigen complexes, in *Catalytic Antibodies*, D. J. Chadwick and J. Marsh (eds.), Wiley, Chichester, UK, 1991; Ciba Foundation Symp. 159, pp. 13–39.
60. Tooze, B. C., *Introduction to Protein Structure*, Garland Publishing, New York, 1991.

61. Kubitz, D. and Keinan, E., Production of monoclonal catalytic antibodies: Principles and practice, in *Catalytic Antibodies*, E. Keinan (ed.), Wiley-VCH, Weinheim, 2005; pp. 491–504.

62. Burton, D. R., Antibody libraries, in *Phage Display: A Laboratory Manual*, C. F. Barbas III, D. R. Burton, J. K. Scott, and G. J. Silverman (eds.), Cold Spring Harbor Laboratory Press, New York, 2001, pp. 3.1–3.18.

63. Marks, J. D., Hoogenboom, H. R., Griffiths, A. D., and Winter, G., Molecular evolution of proteins on filamentous phage. Mimicking the strategy of the immune system, *J. Biol. Chem.* **267**(23):16007–16010 (1992).

64. Mills, D. R., Peterson, R. L., and Spiegelman, S., An extracellular Darwinian experiment with a self-duplicating nucleic acid molecule, *Proc. Natl. Acad. Sci. USA* **58**:217–224 (1967).

65. Spiegelman, S., Experimental analysis of precellular evolution, *Quart. Rev. Biophys.* **4**:213–253 (1971).

66. Powell, K. A., Ramer, S. W., del Cardayré, S. B., Stemmer, W. P. C., Tobin, M. B., Longchamp, P. F., and Huisman, G. W., Directed evolution and biocatalysis, *Angew. Chem. Int. Ed.* **40**(21):3948–3959 (2001).

67. Stemmer, W. and Holland, B., Survial of the fittest molecule, *Am. Sci.* **91**:526–533 (2003).

68. Brik, A., Dawson, P. E., and Keinan, E., The product of the natural reaction catalyzed by 4-oxalocrotonate tautomerase becomes an affinity label of its mutant, *Bioorg. Med. Chem.* **10**:3891–3917 (2002).

69. Piran, R. and Keinan, E., In vitro evolution of catalytic antibodies and other proteins via combinatorial libraries, in *Catalytic Antibodies*, E. Keinan (ed.), Wiley-VCH, Weinheim, 2005, pp. 243–283.

70. Holland, J. H., *Adaptation in Natural and Artificial Systems*, MIT Press, Cambridge, MA, 1992.

71. You, L. and Arnold, F. H., Directed evolution of subtilisin E in Bacillus subtilis to enhance total activity in aqueous dimethylformamide, *Protein Eng.* **9**:77–83 (1996).

72. Hilvert, D., Critical analysis of antibody catalysis, in *Catalytic Antibodies*, E. Keinan (ed.), Wiley-VCH, Weinheim, 2005, pp. 30–71.

73. Otten, L. G. and Quax, W. J., Directed evolution: Selecting today's biocatalysts, *Biomol. Eng.* **22**(1–3):1–9 (2005).

74. Stemmer, W. P. C., Rapid evolution of a protein in vitro by DNA shuffling, *Nature* **370**(6488):389–391 (1994).

75. Stemmer, W. P. C., DNA shuffling by random fragmentation and reassembly: In vitro recombination for molecular evolution, *Proc. Natl. Acad. Sci. USA* **91**(22):10747–10751 (1994).

76. Crameri, A., Raillard, S.-A., Bermudez, E., and Stemmer, W. P. C., DNA shuffling of a family of genes from diverse species accelerates directed evolution, *Nature* **391**(6664):288–291 (1998).

77. Kikuchi, M., Ohnishi, K., and Harayama, S., Novel family shuffling methods for the in vitro evolution of enzymes, *Gene* **236**(1):159–167 (1999).

78. Kikuchi, M., Ohnishi, K., and Harayama, S., An effective family shuffling method using single-stranded DNA, *Gene* **243**(1–2):133–137 (2000).

79. Lorimer, I. A. J. and Pastan, I., Random recombination of antibody single chain Fv sequences after fragmentation with DNase1 in the presence of Mn^{2+}, *Nucl. Acids Res.* **23**(15):3067–3068 (1995).

80. Miyazaki, K., Random DNA fragmentation with endonuclease V: Application to DNA shuffling, *Nucl. Acids Res.* **30**(24):e139 (2002).

81. Shao, Z., Zhao, H., Giver, L., and Arnold, F. H., Random-priming in vitro recombination: An effective tool for directed evolution, *Nucl. Acids Res.* **26**(2):681–683 (1998).

82. Lee, S. H., Ryu, E. J., Kang, M. J., Wang, E.-S., Piao, Z. B., Choi, Y. J., Jung, K. H., Jeon, J. Y. J., and Shin, Y. C., A new approach to directed gene evolution by recombined extension on truncated templates (RETT), *J. Mol. Catal. B: Enzymatic* **26**(3–6):119–129 (2003).

83. Lutz, S., Ostermeier, M., Moore, G. L., Maranas, C. D., and Benkovic, S. J., Creating multiple-crossover DNA libraries independent of sequence identity, *Proc. Natl. Acad. Sci. USA* **98**(20):11248–11253 (2001).

84. Zhao, H., Giver, L., Shao, Z., Affholter, J. A., and Arnold, F. H., Molecular evolution by staggered extension process (StEP) in vitro recombination, *Nature Biotechnol.* **16**(3):258–261 (1998).

85. Coco, W. M., Levinson, W. E., Crist, M. J., Hektor, H. J., Darzins, A., Pienkos, P. T., Squires, C. H., and Monticello, D. J., DNA shuffling method for generating highly recombined genes and evolved enzymes, *Nature Biotechnol.* **19**(4):354–359 (2001).

86. Ostermeier, M., Shim, J. H., and Benkovic, S. J., A combinatorial approach to hybrid enzymes independent of DNA homology, *Nature Biotechnol.* **17**(12):1205–1209 (1999).

87. Lutz, S., Ostermeier, M., and Benkovic, S. J., Rapid generation of incremental truncation libraries for protein engineering using α-phosphothioate nucleotides, *Nucl. Acids Res.* **29**:16 (2001).

88. Sieber, V., Martinez, C. A., and Arnold, F. H., Libraries of hybrid proteins from distantly related sequences, *Nature Biotechnol.* **19**:456–460 (2001).

89. Kolkman, J. A. and Stemmer, W. P. C., Directed evolution of proteins by exon shuffling, *Nature Biotechnol.* **19**(5):423–428 (2001).

90. Gibbs, M. D., Nevalainen, K. M. H., and Bergquist, P. L., Degenerate oligonucleotide gene shuffling (DOGS): A method for enhancing the frequency of recombination with family shuffling, *Gene* **271**(1):13–20 (2001).

91. Hiraga, K. and Arnold, F. H., General method for sequence-independent site-directed chimeragenesis, *J. Mol. Biol.* **330**(2):287–296 (2003).

92. Kitamura, K., Kinoshita, Y., Narasaki, S., Nemoto, N., Husimi, Y., and Nishigaki, K., Construction of block-shuffled libraries of DNA for evolutionary protein engineering: Y-ligation-based block shuffling, *Protein Eng.* **15**(10):843–853 (2003).

93. O'Maille, P. E., Bakhtina, M., and Tsai, M.-D., Structure-based combinatorial protein engineering (SCOPE). *J. Mol. Biol.* **321**(4):677–691 (2002).

94. Abecassis, V., Pompon, D., and Truan, G., High efficiency family shuffling based on multi-step PCR and in vivo DNA recombination in yeast: Statistical and functional analysis of a combinatorial library between human cytochrome P450 1A1 and 1A2, *Nucl. Acids Res.* **28**:E88 (2000).

95. Strausberg, S. L., Alexander, P. A., Gallagher, D. T., Gilliland, G. L., Barnett, B. L., and Bryan, P. N., Directed evolution of a subtilisin with calcium-independent stability, *Bio/Technology* **13**(7):669–773 (1995).

96. Stemmer, W. P. C., Crameri, A., Ha, K. D., Brennan, T. M., and Heyneker, H. L., Single-step assembly of a gene and entire plasmid from large numbers of oligodeoxyribonucleotides, *Gene* **164**(1):46–53 (1995).

97. Coco, W. M., Encell, L. P., Levinson, W. E., Crist, M. J., Loomis, A. K., Licato, L. L., Arensdorf, J. J., Sica, N., Pienkos, P. T., and Monticello, D. J., Growth factor engineering by degenerate homoduplex gene family recombination, *Nature Biotechnol.* **20**(12):1250 (2002).

98. Ness, J. E., Kim, S., Gottman, A., Pak, R., Krebber, A., Borchert, T. V., Govindarajan, S., Mundorff, E. C., and Minshull, J., Synthetic shuffling expands functional protein diversity by allowing amino acids to recombine independently, *Nature Biotechnol.* **20**(12):1251–1255 (2002).

99. Zha, D., Eipper, A., and Reetz, M. T., Assembly of designed oligonucleotides as an efficient method for gene recombination: A new tool in directed evolution, *ChemBioChem* **4**(1):34–39 (2003).

100. Chen, K. and Arnold, F. H., Tuning the activity of an enzyme for unusual environments: Sequential random mutagenesis of subtilisin E for catalysis in dimethylformamide, *Proc. Natl. Acad. Sci. USA* **90**(12):5618–5622 (1993).

101. Arnold, F. H., Design by directed evolution, *Acc. Chem. Res.* **31**:125–131 (1998).

102. Reetz, M. T., Combinatorial and evolution-based methods in the creation of enantioselective catalysts, *Angew. Chem. Int. Ed.* **40**:284–310 (2001).

103. Reetz, M. T., Directed evolution as a means to create enantioselective enzymes for use in organic chemistry, in *Directed Molecular Evolution of Proteins*, S. Brakmann and K. Johnsson (eds.), Wiley-VCH, Weinheim, 2002, pp. 245–279.

104. Sambrook, J., Fritsch, E. F., and Maniatis, T., *Molecular Cloning: A Laboratory Manual*, Cold Spring Harbor Laboratory Press, New York, 1989, Vols. 1–3.

105. Cleland, J. L., Jones, A. J. S., and Craik, C. S., Introduction to protein engineering, in *Protein Engineering: Principles and Practice*, J. L. Cleland and C. S. Craik (eds.), Wiley-Liss, New York, 1996, pp. 1–32.

106. Moore, J. C. and Arnold, F. H., Directed evolution of a para-nitrobenzyl esterase for aqueous-organic solvents, *Nature Biotechnol.* **14**:458–467 (1996).

107. Reetz, M. T., Controlling the enantioselectivity of enzymes by directed evolution: Practical and theoretical ramifications, *Proc. Natl. Acad. Sci. USA* **101**(16):5716–5722 (2004).

108. Reetz, M. T., Zonta, A., Schimossek, K., Jaeger, K.-E., and Liebeton, K., Creation of enantioselective biocatalysts for organic chemistry by in vitro evolution, *Angew. Chem. Int. Ed.* **36**(24):2830–2832 (1997).

109. Reetz, M. T. and Jaeger, K.-E., Enantioselective enzymes for organic synthesis created by directed evolution, *Chem. Eur. J.* **6**:407–412 (2000).

110. Zha, D., Wilensek, S., Hermes, M., Jaeger, K.-E., and Reetz, M. T., Complete reversal of enantioselectivity of an enzyme-catalyzed reaction by directed evolution, *Chem. Commun.* **2001**(24):2664–2665 (2001).

111. Smith, G. P. and Petrenko, V. A., Phage display, *Chem. Rev.* **97**:391–410 (1997).
112. Model, P. and Russel, M., Filamentous bacteriophage, in *The Bacteriophages*, R. Calendar (ed.), Plenum Press, New York, 1988, Vol. 2, pp. 375–456.
113. Webster, R. E. and Lopez, J., Structure and assembly of the class 1 filamentous bacteriophage, in *Virus Structure and Assembly*, S. Casjens (ed.), Jones & Bartlett, Boston, 1985, pp. 235–268.
114. Castagnoli, L., Zuccconi, A., Quondam, M., Rossi, M., Vaccaro, P., Panni, S., Paoluzi, S., Santonico, E., Dente, L., and Cesareni, G., Alternative bacteriophage display systems, *Combin. Chem. High Throughput Screen.* **4**(2):121–133 (2001).
115. Verhaert, R. M. D., van Duin, J., and Quax, W. J., Processing and functional display of the 86 kDa heterodimeric penicillin G acylase on the surface of phage fd, *Biochem. J.* **342**(2):415–422 (1999).
116. Wentworth, Jr., P., Antibody design by man and nature, *Science* **296**(5576):2247–2249 (2002).
117. Soumillion, P. and Fastrez, J., Investigation of phage display for the directed evolution of enzymes, in *Directed Molecular Evolution of Proteins*, S. Brakmann and K. Johnsson (eds.), Wiley-VCH, Weinheim, 2002, pp. 79–110.
118. Baca, M., Scanlan, T. S., Stephenson, R. C., and Wells, J. A., Phage display of a catalytic antibody to optimize affinity for transition-state analog binding, *Proc. Natl. Acad. Sci. USA* **94**:10063–10068 (1997).
119. Cesaro-Tadic, S., Lagos, D., Honegger, A., Rickard, J. H., Partridge, L. J., Blackburn, G. M., and Plückthun, A., Turnover-based in vitro selection and evolution of biocatalysts from a fully synthetic antibody library, *Nature Biotechnol.* **21**(6):679–685 (2003).
120. Kruger, K., Grabowski, P. J., Zaug, A. J., Sands, J., Gottschling, D. E., and Cech, T. R., Self-splicing RNA: Autoexcision and autocyclization of the ribosomal RNA intervening sequence of Tetrahymena, *Cell* **31**:147–157 (1982).
121. Zaug, A. J. and Cech, T. R., The intervening sequence RNA of Tetrahymena is an enzyme, *Science* **231**:470–475 (1986).
122. Guerrier-Takada, C., Gardiner, K., Marsh, T., Pace, N., and Altman, S., The RNA moiety of ribonuclease P is the catalytic subunit of the enzyme, *Cell* **35**:849–857 (1983).
123. Breaker, R. R., In vitro selection of catalytic polynucleotides, *Chem. Rev.* **97**:371–390 (1997).
124. Robertson, D. L. and Joyce, G. F., Selection in vitro of an RNA enzyme that specifically cleaves single-stranded DNA, *Nature* **344**(6265):367–368 (1990).
125. Ellington, A. D. and Szostak, J. W., In vitro selection of RNA molecules that bind specific ligands, *Nature* **346**(6287):818–822 (1990).
126. Tuerk, C. and Gold, L., Systematic evolution of ligands by exponential enrichment: RNA ligands to bacteriophage T4 DNA polymerase, *Science* **249**(4968):505–510 (1990).
127. McPherson, M. J. and Moller, S. G., *PCR*, Springer, New York, 2000.
128. Bartel, D. P. and Szostak, J. W., Isolation of new ribozymes from a large pool of random sequences, *Science* **261**(5127):1411–1418 (1993).

129. Hermanson, G. T., *Bioconjugate Techniques*, Academic Press, San Diego, 1996.

130. Haller, A. A. and Sarnow, P., In vitro selection of a 7-methyl-guanosine binding RNA that inhibits translation of capped mRNA molecules, *Proc. Natl. Acad. Sci. USA* **94**(16):8521–8526 (1997).

131. Wilson, D. S. and Szostak, J. W., In vitro selection of functional nucleic acids, *Annu. Rev. Biochem.* **68**:611–647 (1999).

132. Lorsch, J. R. and Szostak, J. W., Chance and necessity in the selection of nucleic acid catalysts, *Acc. Chem. Res.* **29**:103–110 (1996).

133. Faulhammer, D. and Famulok, M., The Ca^{2+} ion as a cofactor for a novel RNA-cleaving deoxyribozyme, *Angew. Chem. Int. Ed.* **35**:2837–2841 (1996).

134. Roth, A. and Breaker, R. R., An amino acid as a cofactor for a catalytic polynucleotide, *Proc. Natl. Acad. Sci. USA* **95**(11):6027–6031 (1998).

135. Conn, M. M., Prudent, J. R., and Schultz, P. G., Porphyrin metallation catalyzed by a small RNA molecule, *J. Am. Chem. Soc.* **118**:7012–7013 (1996).

136. Li, Y. and Sen, D., A catalytic DNA for porphyrin metallation, *Nature Struct. Biol.* **3**:743–747 (1996).

137. Fersht, A., *Enzyme Structure and Mechanism*, 2nd ed.; Freeman, New York, 1985.

138. Silverman, S. K., Deoxyribozymes: DNA catalysts for bioorganic chemistry, *Org. Biomol. Chem.* **2**:2701–2706 (2004).

139. Watson, J. D. and Crick, F. H. C., Molecular structure of nucleic acids, *Nature* **171**:737–738 (1953).

140. Emilsson, G. M. and Breaker, R. R., Deoxyribozymes: New activities and new applications, *Cell. Mol. Life Sci.* **59**:596–607 (2002).

141. Peracchi, A., DNA catalysis: Potential, limitations, open questions, *ChemBioChem* **6**:1316–1322 (2005).

142. Carmi, N., Shultz, L. A., and Breaker, R. R., In vitro selection of self-cleaving DNAs, *Chem. Biol.* **3**(12):1039–1046 (1996).

143. Lu, Y., New transition-metal-dependent DNAzymes as efficient endonucleases and as selective metal biosensors, *Chem. Eur. J.* **8**(20):4588–4596 (2002).

144. Breaker, R. R., Natural and engineered nucleic acids as tools to explore biology, *Nature* **432**:838–845 (2004).

145. Seetharaman, S., Zivarts, M., Sudarsan, N., and Breaker, R. R., Immobilized RNA switches for the analysis of complex chemical and biological mixtures, *Nature Biotechnol.* **19**(4):336–341 (2001).

146. Larhed, M. and Hallbert, A., The Heck reaction and related carbopallidation reactions, in *Handbook of Organopalladium Chemistry for Organic Synthesis*, E.-I. Negishi (ed.), Wiley, New York, 2002, Vol. 1, pp. 1135–1178.

147. Vannela, R. and Woo, L. K., Development of homogeneous catalysts using molecular evolution, Abstract of Papers, 232nd American Chemical Society National Meeting, San Francisco, CA, Sept. 10–14, 2006 (2006), INOR-087.

148. Jaschke, A. and Seelig, B., Evolution of DNA and RNA as catalysts for chemical reactions, *Curr. Opin. Chem. Biol.* **4**(3):257–262 (2000).

149. Joyce, G. F., Nucleic acid enzymes: Playing with a fuller deck, *Proc. Natl. Acad. Sci. USA* **95**:5845–5847 (1998).

150. Kool, E. T., Replacing the nucleobases in DNA with designer molecules, *Acc. Chem. Res.* **35**:936–943 (2002).

151. Yu, C., Henry, A. A., Romesberg, F. E., and Schultz, P. G., Polymerase recognition of unnatural base pairs, *Angew. Chem. Int. Ed.* **41**(20):3841–3844 (2002).

152. Tae, E. L., Wu, Y., Xia, G., Schultz, P. G., and Romesberg, F. E., Efforts toward expansion of the genetic alphabet: Replication of DNA with three base pairs, *J. Am. Chem. Soc.* **123**(30):7439–7440 (2001).

153. Johns, G. C. and Joyce, G. F., The promise and peril of continuous in vitro evolution, *J. Mol. Evol.* **61**:253–263 (2005).

CHAPTER 7

Biomaterials Informatics

NICOLE K. HARRIS and JOACHIM KOHN
Department of Chemistry and Chemical Biology
Rutgers University
Piscataway, New Jersey

WILLIAM J. WELSH
The Informatics Institute
University of Medicine and Dentistry of New Jersey
Newark, New Jersey

DOYLE D. KNIGHT
Department of Mechanical and Aerospace Engineering
Rutgers University
Piscataway, New Jersey

7.1. INTRODUCTION

The first modern biomaterials appeared in the mid-20th century and were utilized for hip joint replacement, vascular prostheses, heart valves, and eye lens implants [1]. These devices were designed to be inert (nondegradable) and replace defective tissue and/or bone [2]. These biomaterials were typically adopted from materials developed for other scientific applications. For example, poly(ether urethane)s were initially used to make ladies' girdles and later in artificial hearts [3]. While these materials helped realize new and essential treatments, they were not optimized for medical use. These biostable, synthetic implant materials lacked the molecular sequences and patterns crucial to normal cell function and often triggered aberrant cell responses on long-term implantation [4].

New biodegradable polymeric materials that mimic the properties of body tissues are needed for biomedical applications [5,6]. The current set

Combinatorial Materials Science, Edited by Balaji Narasimhan, Surya K. Mallapragada, and Marc D. Porter

of approved synthetic degradable polymers, however, is very limited. Since 1969, polymers derived from lactic and glycolic acid [7,8], polydioxanone [9], and a polyanhydride [10] derived from sebacic acid and bis(*p*-carboxyphenoxy)propane have been the only synthetic degradable polymers that have an extensive regulatory approval history in the United States.

The potential number of different biodegradable polymers for biomedical applications is vast. For example, there are at least 12 different families of biodegradable polymers potentially useful as tissue regeneration scaffolds (see Table 7.1 [11]). The number of possible structural variations of these materials has been estimated by Kohn (unpublished communication) to exceed 20,000 discrete compositions, each one having a different mix of physicome-

TABLE 7.1. Synthetic Degradable Polymers[a] and Representative Applications

Library No.	Category	Current Major Research Applications
1	Polyarylates	Cartilage repair, nerve repair, coatings for medical devices
2	Poly(glycolic acid), poly(lactic acid), and copolymers	Barrier membranes, drug delivery, guided tissue regeneration (in dental applications), orthopedic applications, stents, staples, sutures, tissue engineering
3	Polyhydroxybutyrate (PHB), polyhydroxyvalerate (PHV), and copolymers thereof	Long-term drug delivery, orthopedic applications, stents, sutures
4	Polycaprolactone	Long-term drug delivery, orthopedic applications, staples, stents
5	Polydioxanone	Fracture fixation in non-load-bearing bones, sutures, wound clip
6	Polyanhydrides	Drug delivery
7	Polycyanoacrylates	Adhesives, drug delivery, wound closure
8	Poly(amino acids) and "pseudo"-poly(amino acids)	Drug delivery, tissue engineering, orthopedic applications
9	Poly(ortho ester)	Drug delivery, stents
10	Polyphosphazenes	Blood-contacting devices, drug delivery, sceletal reconstruction
11	poly(propylene fumarate)	Orthopedic applications
12	Tyrosine-derived polycarbonates and related materials	Cardiovascular applications, stents, bone fixation, drug delivery

[a] Estimated total number of polymers = 20,000.

Source: Ref. 11.

chanical and biological properties. This compendium of biodegradable polymers far exceeds the number of specimens that can reasonably be evaluated and tested in vitro (much less in vivo) for specific biomedical applications.

The development of new biomaterials for biomedical applications therefore requires a new research paradigm for effectively and efficiently screening the vast libraries of potential polymers to determine the optimal molecular structures. The new research paradigm is a synergy of three technologies: combinatorial analysis and parallel synthesis, rapid screening, and computational modeling. In this chapter, we describe recent advances in all three areas, and illustrate the dramatic progress that has been achieved by this approach.

The chapter is organized as follows. Section 7.2 provides a background to combinatorial analysis. Section 7.3 describes the tyrosine-based polymer libraries developed in our research. Section 7.4 discusses combinatorial materials analysis. Section 7.5 describes biochemical assays and rapid screening. Section 7.6 presents the computational models. Section VII discusses the conclusions and future prospects.

7.2. BACKGROUND

Correlations between chemical structure and the properties of polymers have been explored since the early 1930s, when the macromolecular structure of polymers was first recognized. Determining the tensile strength of glass, iron, paper, wood, and polyethylene may lead to the identification of a material with suitable strength for a given application, but determining the effect of systematic changes in structure on the tensile properties in a set of polymers that share some common structural features is a more rewarding approach that provides a method to develop general structure–property relationships. Unfortunately, the correlations are often limited to the test set of polymers, and complex correlations between biological properties and chemical structure cannot be determined. Larger sets of polymers, efficient characterization techniques, and computational models are often required to capture these structure–property relationships.

While many researchers now adopt the language of combinatorial chemistry and refer to their selection of test materials as a "library," we attempt to be rigorous in distinguishing between "sets of test materials" and a true "library," which we define as a group of test materials that share common properties and structural features. These shared properties and features facilitate the identification of powerful structure–property–performance correlations that can be extrapolated to thousands of polymer structures—beyond the limited number of specific polymers used in establishing the correlations. This represents the *added value* associated with the use of combinatorial biomaterial libraries.

A combinatorial approach is most effective when clear correlations between the basic design variables (e.g., biomaterial chemistry and structure) and the performance of the product (e.g., cell–biomaterial interactions) are not available. Obviously, the development of synthetic biomaterials for use in tissue engineering or drug delivery falls into this category. One consequence of the proposed combinatorial synthetic schemes is a dramatic increase in the number of polymers available for exploration. The increased volume of products requires the development of rapid and efficient characterization technologies, and computational strategies.

While the idea of extending combinatorial chemistry from drug discovery to materials research may seem straightforward, there are some fundamental differences that should be clarified. First, the goal of the combinatorial approach in drug discovery is to find "lead" compounds, or specific compounds that exhibit specific bioactivity in a specific bioassay, out of a very large library. In drug discovery, the binding site or molecule of interest is predefined for the bioassay and small quantities of material are sufficient for initial screening. These initial hits are used to form the framework of the core design(s), which are then further investigated by other scientists. On the other hand, the complex requirements of materials for a specific biomedical application cannot be captured in a single bioassay. Therefore, the basic approach to combinatorial chemistry of biomaterials design requires significant conceptual and practical modification of the basic paradigm in drug discovery. Second, the complicated interplay of materials properties and biocompatibility results in technical problems and challenges. Specifically, parallel synthesis of polymers is more complex than for small molecules. Polymer properties depend on molecular weight and processing conditions, and biological properties are also influenced by the shape of the device and surface properties. This list is rather daunting.

The goal of combinatorial methods in biomaterials research is focused on the thorough investigation of small libraries of structurally related polymers in order to generate datasets for development, analysis, and prediction of complex structure–property relationships. Variables can be minimized by choosing alternating copolymers over random copolymers, choosing monomers that are polymerized under identical reaction conditions to give polymers of similar molecular weight and molecular weight distribution, isolating the polymer in pure form, and processing all polymer samples identically. In the case of biomaterials research, each material provides useful information and larger quantities of purified materials are required to complete the characterization of the material, ~40–100 mg.

The combinatorial approach to biomaterials is comprised of three key components: parallel synthesis, high-throughput characterization or rapid screening, and computational modeling. Ultimately, the goal is to develop comprehensive datasets ranging from physicomechanical properties, surface chemistry, protein adsorption, cellular responses in vitro to tissue responses in vivo. Appropriate analysis of these data using modeling approaches will hopefully result in structure–property–performance correlations that enable prediction of the "biological performance" of a material *in vitro* and/or *in vivo*, and allow for implant design optimization.

This basic methodology has also been successfully implemented in the design of catalytically active polymers [12], dental materials [13], and coating technologies [14]. Other polymerization strategies, including atom transfer free-radical polymerization and ring-opening polymerization, have recently been demonstrated using automated synthesis [15]. Ultimately, promising new biomaterials will emerge from this research effort for use in a wide range of tissue engineering and drug/gene delivery applications.

7.3. TYROSINE-BASED POLYMERS AS BIOMATERIALS

7.3.1. Tyrosine-Derived Diphenols as Monomers

The polymers discussed in this chapter were synthesized from a nontoxic, biocompatible diphenolic monomer, desaminotyrosyl–tyrosine alkyl ester (DTR; Fig. 7.1), which is a derivative of naturally occurring tyrosine dipeptide with the important structural modification that the *N* terminus of the peptide is replaced by a hydrogen atom and the *C* terminus of the peptide is protected by an alkyl ester chain of variable length and structure. The major structural features of the DTR unit include (1) two stiff phenyl rings that result in polymers with excellent mechanical properties; (2) appreciable hydrophobicity, which can be modulated by changing the length of the alkyl ester pendent chain; (3) the presence of a peptide bond between desaminotyrosine and tyrosine, which facilitates hydrogen bonding interactions; and (4) the presence

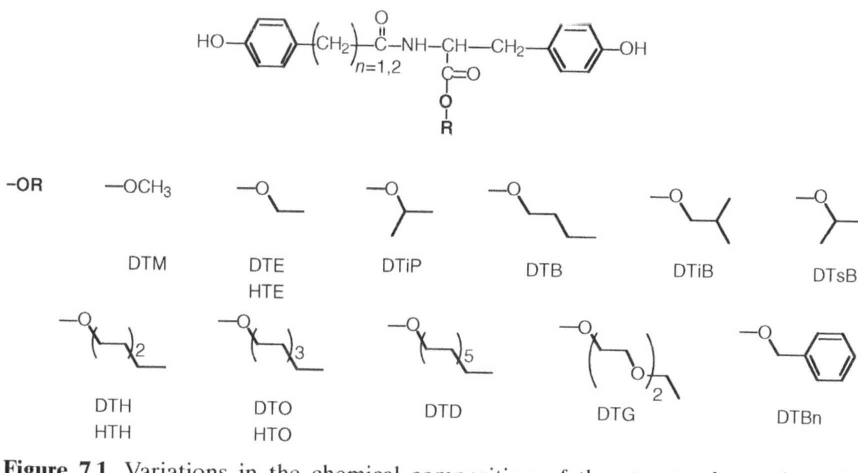

Figure 7.1. Variations in the chemical composition of the structural template of a tyrosine-derived diphenol (monomer A). HTR when $n = 1$, DTR when $n = 2$.

of an optically active carbon atom on the L-tyrosine unit. In combination, *these structural features are unique among all currently used polymeric biomaterials* and provide the basis for some of the outstanding material properties observed in DTR-containing polymers. Polycarbonates derived from DTR combine the good mechanical properties of industrial polycarbonates (used in bulletproof glass, in high-quality kitchen utensils, and in shatterproof lenses) with the nontoxicity and high degree of biocompatibility inherent in a polymer whose sole degradation products are naturally occurring metabolites [16]. In particular, DTE has passed all biocompatibility tests required by the FDA, and poly(DTE carbonate) is currently under FDA review for use in several orthopedic implants.

7.3.2. Polyarylate Library

The basic approach consists of the use of strictly alternating AB copolymers derived from a set of *x* structural variations of A and *y* structural variations of B. The copolymerization of all the variations of A with all the variations of B in all possible combinations results in *x* times *y* individual polymer compositions. We have used this approach to create a combinatorial library of 112 degradable polyarylates based on the copolymerization of 14 tyrosine-derived diphenols (the A monomer) with 8 distinct diacids (the B monomer) in all possible combinations as indicated in Figures 7.2 and 7.3 [17–19]. The diacids include various nutrients, constituents of the Krebs cycle, and food additives approved by the FDA (EAFUS listing). The structural variation in the side chain of the DTR monomer and in the backbone of the diacids produced an

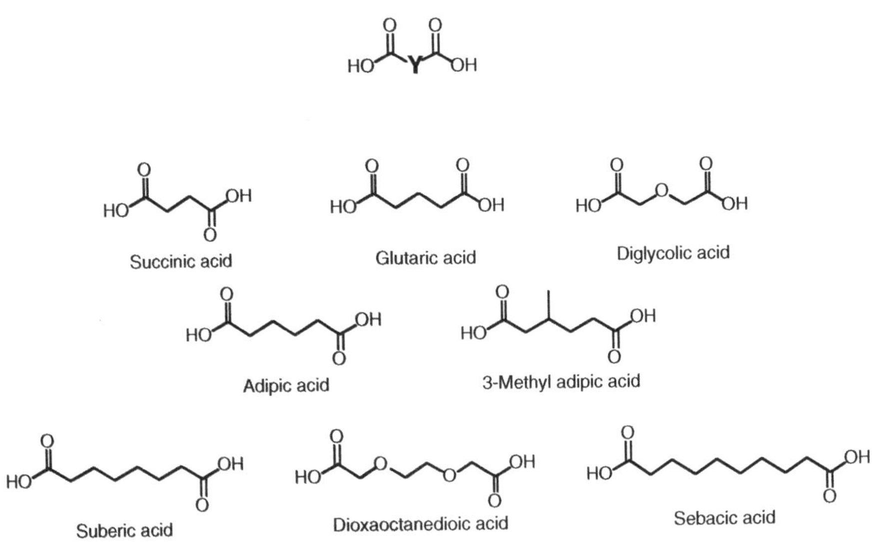

Figure 7.2. Variations in the chemical composition of the diacids used in this library (monomer B).

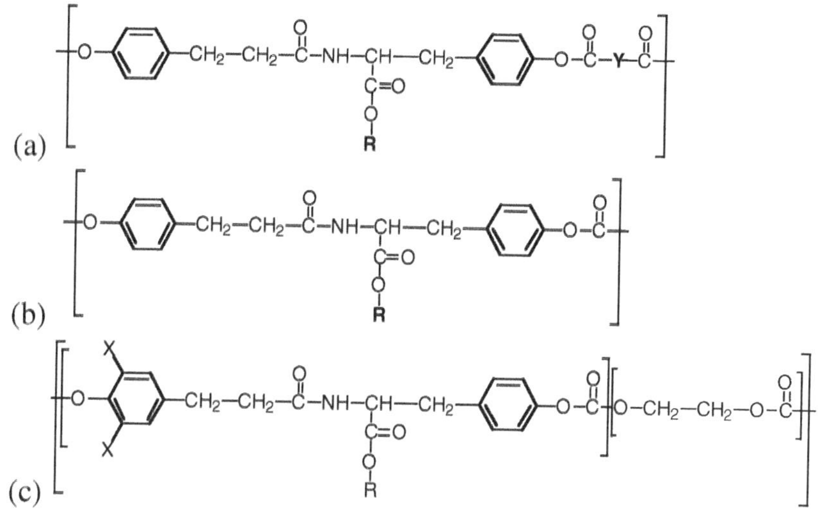

Figure 7.3. (a) Polyarylate R = alkyl esters, Y = various diacids; (b) poly(DTR carbonate); (c) poly((X_2–DTR-co-*y*%(X_2–DT)-co-*z*%(PEG_{nK}) carbonate) X = H,I; R = H, ethyl,butyl,hexyl,octyl,dodecyl.

orthogonal library, since the backbone and side chain of the polymer elicit different properties. Additive libraries, or libraries where both monomers can only vary in one direction (i.e., both monomers have different number of methylene spacers in the backbone) do not offer any unique structure–function information. The synthesis and characterization of this library resulted in correlations between polymer structure and their physicomechanical properties and various cellular responses in vitro [17–19].

Each A monomer was polymerized with each B monomer using room-temperature polyesterification [20] in a parallel reaction setup. The polymers were isolated and purified by repeated precipitation within the original reaction vessels, providing 40–50 mg of 112 individual polymers. The polymers obtained had similar molecular weights and molecular weight distributions. Molecular, physical, and biological properties were characterized for each sample. Differential scanning calorimetry (DSC) revealed that the polymers range from amorphous to liquid crystalline, span a broad range of glass transition temperatures (~100°C), and exhibit high thermal stability up to 300°C. The material properties of these polymers were also very diverse, ranging from soft, elastomeric to hard, tough materials. The air–water contact angle was determined using goniometry for each of the polyarylates, with values ranging from 64° to 101°. While this polymer library covered a substantial range of thermal, mechanical, and surface properties, all of these polymers were water-insoluble, thermoplastic materials with potential application as an implant device or tissue regeneration scaffold [17]. The biological assays are discussed in detail later in this chapter.

7.3.3. Polycarbonates

Random copolymers of DTR monomer and PEG were synthesized by condensation polymerization with phosgene, providing polymers with molecular weights ranging from 40,000 to 200,000 as indicated in Figure 7.3 [21]. Nearly a dozen different copolymers were prepared by varying the DTR monomer, the molecular weight of PEG, and by varying the ratio of DTR to PEG. Characterization of the molecular, physicomechanical, and biological properties resulted in the development of the following structure-property relationships. PEG content and length of the DTR sidechain were inversely correlated with the glass transition temperature and tensile modulus; however, increasing the length of the PEG block resulted in a higher glass transition temperature and higher tensile modulus compared to copolymers containing low-molecular-weight PEG. Increasing PEG content in the copolymers also increased water uptake and the rate of backbone degradation, but inhibited cell adhesion and proliferation.

An in-depth investigation on the effect of incorporating PEG (1000 g/mol, 0–10 mol%) into poly(DTE carbonate) by copolymerization revealed that the presence of PEG clearly had a pronounced effect on the adsorption of serum proteins on the polymer surfaces [22] and affected fundamental cellular responses such as attachment, migration, and proliferation in a way that would not have been predicted. Interestingly, the PEG-containing copolymers adsorb fibronectin, a cell-binding ligand, which seems to induce motility of human primary epidermal cells, keratinocytes [22,23]. Briefly, cell migrations rates were similar on substrates comprised of 4 and 8 mol% PEG, despite a decrease in the number of available cell-binding domains. This effect was explained by cellular remodeling of the surface containing 8 mol% PEG to increase the number of available binding domains. While previous studies of biomaterials have supported that the incorporation of PEG dramatically reduces cell and protein adhesion [24], it is possible that there is a low-PEG concentration regime whereby cells and proteins are still able to attach and proliferate [23]. This in vitro result suggests that these materials may be useful in wound-healing therapies by facilitating the natural healing response: epidermal cells migrate to clear and remodel the wounded area.

Rational design was used for the development of a radioopaque biodegradable stent [25–27]. First, a virtual library was designed according to the requirements specific to the stent application. Poly(DTE carbonate) was chosen as the base of the material since it degraded slowly under physiological conditions and exhibited excellent biocompatibility [21,28,29]. Then, as it is important that stents be antithrombogenic, or that they deter clot formation around the device, polyethylene glycol (PEG) was selected as a component since the incorporation of PEG into poly(DTE carbonate) reduced protein adsorption and the stiffness of the parent biomaterial [21,30]. The resorption of the stent and drug delivery was designed to be tuned by the incorporation of free carboxylic acids into the polymer backbone, by means of desaminotyrosyl–tyrosine monomer (DT) [26,31]. Finally, an added advantage of this material over current polymer-based options was the introduction of the ability to view the integrity and stability of the stent in a noninvasive manner using standard X-ray imaging [32]. The method of introducing radioopacity into the polymer was accomplished by the incorporation of iodine into the tyrosine-based polymer repeat units. Similar iodinated compounds have been synthesized for use as X-ray contrast agents [33], and have been identified in the thyroid gland, β-[3,5-diiodo-4-hydroxyphenoxy)-3,5-diiodophenyl]alanine (thyroxine) [34], which is synthesized from 3,5-diiodotyrosine [35]. The resulting polymer has the following general formula: poly(I_xDTR-co-y%(I_xDT)-co-z%(PEG_{nK}) carbonate) (Fig. 7.3). The stiffness of the material was also affected by these different components; the incorporation of iodine or DT increased the stiffness, while the incorporation of PEG or longer chain tyrosine esters decreased the stiffness [28]. While this library was created with rational design, we were not sure how the biological properties or the degradation profile would be affected by the incorporation of PEG, iodine, and DT; however, the property profile can be finely tuned by varying the molecular weight of the PEG block (n), the length of the DTR sidechain (R), the ratio of the three components, and the level of iodination (x), easily resulting in 10,000 unique combinations. As expected, initial results revealed that the incorporation of 15 mol% PEG into poly(I_2DTE carbonate) cut protein adsorption by nearly 95% relative to iodinated and non-iodinated poly(DTE carbonate). While the incorporation of iodine slightly increased the air–water contact angle compared to the parent biomaterial, the level of protein adsorption was essentially the same [36].

7.4. COMBINATORIAL BIOMATERIALS ANALYSIS

The challenge of evaluating combinatorial libraries of biomaterials is twofold: (1), the relationship between the structure of the material and its biological response has not been elucidated to enable the development of appropriate rapid screening of libraries using simple bioassays; and (2) the performance of the materials is dependent on a number of often interrelated factors, including molecular weight, hydrophobicity, stiffness, and biodegradability. Furthermore, the method of polymer processing can produce two distinctly different materials from the same polymer. Consider the properties of a foam coffee cup versus a overhead transparency or a CD case. They are all made of polystyrene, yet each is distinctly unique [6]. To meet these challenges, innovative rapid screening tools, which are inexpensive and can be utilized in parallel, have been developed to characterize the biological response of a family of structurally–related materials. The variables affecting polymer properties were reduced by choosing polymers with similar molecular weights and molecular weight distributions, which were processed in the same manner. Since the goal of combinatorial materials research is to learn as much as possible about each material, a number of characterization techniques that require a minimal amount of sample were performed, including ^{1}H NMR spectroscopy, size exclusion chromatography (SEC), DSC, and goniometry. Furthermore, tensile properties were obtained from miniaturized samples (5 × 50 mm, ~0.01 mm thickness) [18].

7.4.1. Rationale for Dataset Development

The most comprehensive dataset currently available to us contains complete polymer characterization data and a full complement of cell response data for 62 of the 112 members of the polyarylate library [17–19]. This dataset was used in the development of the artificial neural network (ANN) that predicted cell metabolic activity. For the definition of library spaces and for the development of semiempirical models of cell–material interactions, the collection of a wide range of data is *absolutely critical*. It is important to note that we are generating polymer libraries in which the individual members exhibit structure–property correlations that allow one to predict the properties of every member of the library, eliminating the need to actually prepare all possible polymers within the library. Briefly, the artificial neural network described later in this chapter is able not only to predict outcome, but also to determine which parameters among many possible inputs are most closely related to the predicted outcome. Therefore, the surrogate modeling effort can in fact provide mechanistic and fundamental insight—*if* broad ranges of different input parameters are available.

7.4.2. Materials Properties

In most laboratories, characterization of the physicomechanical properties of a biomaterial is generally not high-throughput since much of the instrumentation available is expensive; however, the data acquired using these techniques can be used to screen materials for other important performance properties prior to biological testing. Two measurements that can be made with minimal training and less than 10 mg of sample are (1) the air–water contact angle, a measure of hydrophobicity of the bulk polymer; and (2) the glass transition temperature, the temperature where the amorphous domains of a polymer transition from glassy to rubbery, allowing polymer chains to move. This temperature is closely related to the molecular structure of the material. Both samples may take 5 min to prepare, albeit the samples for thermal analysis are more tedious to prepare, but the contact angle measurement can be made in less than 5 min whereas the thermal analysis takes nearly one hour per sample. The latter characterization method is best suited for a differential scanning calorimetry instrument equipped with an autosampler so samples can run continuously. Alternatively, dynamic mechanical thermal analysis (DMTA) can be used to characterize both the glass transition temperature and the modulus (or stiffness) of the material. While the experiment requires more sample and longer residence times in the instrument (1–4 samples per day), these are both vital parameters to identifying the suitability of a material for a specific application (mechanical and processing properties). In short, the materials scientist requires this information to assess the suitability of a material for a specific application, and if biological data can be predicted based on this standard characterization data, the result is extensive time and money saved on in vitro and in vivo biological assays.

By defining a structural parameter, the *total flexibility index* (TFI) as the total number of carbon atoms found within the *variable* regions of the polymer structure, a simple graphic presentation of glass transition temperatures can be transformed into a correlation between chemical structure in terms of TFI and the glass transition temperature [37]. This correlation resulted in the accurate prediction of the glass transition temperature for all polyarylates in the library, illustrating the advantage of the combinatorial approach over the traditional approach. Likewise, general molecular modeling was used to accurately predict the surface hydrophobicity (as measured by the air–water contact angle) for the entire library [37].

7.4.3. Mass Spectrometry

Thorough characterization of the surface properties of polymer films is essential to establish and understand surface structure–biological property relationships. Time-of-flight secondary-ion mass spectrometry (TOF-SIMS) can be used to identify the polymer repeat units on the surface of the material, which may give specific information about the orientation of block copolymer morphology to the surface (i.e., layered microstructure parallel or orthogonal to the substrate), or simply distinguish a particular polymer from a family of structurally related materials [38]. Characterization of polyarylates using this technique resulted in the identification of fingerprints for both the backbone (or diacid component) and the length of the ester sidechain of the diol component. While TOF-SIMS gives important information about the chemical composition of the monomer, matrix-assisted laser desorption/ionization time-of-flight mass spectroscopy (MALDI-TOF-MS), provides an absolute value for polymer molecular weight, the molecular weight distribution, and the chemical composition of the polymer endgroups. Recently, the development of automated sample preparation for MALDI-TOF-MS was reported [39]. MALDI-TOF MS measurements are made from an evaporated film mixture, containing the polymer solution, a matrix, and a salt additive. Success of the technique depends largely on the appropriate choice of matrix and salt additive. The ability to automate the sample preparation allows for a combinatorial investigation of the appropriate formulation for each polymer, or for a related family of materials.

7.5. BIOCHEMICAL ASSAYS AND RAPID SCREENING

7.5.1. Adaptation of Biochemical Assays for Rapid Screening of Cellular Responses In Vitro

Standard biological assays for the evaluation of a biomaterial for medical applications remain to be developed. Progress is being made to develop, validate, and disseminate the necessary assays, but the process is slow [40–42]. To this date, several conventional tissue culture assays have been successfully adapted to the 96-well plate format. In this way, the results for almost the entire library could be obtained in a reproducible fashion within one plate. Polypropylene were typically used since this material is solvent resistant, allowing for the preparation of polymer films via solvent casting. These assays are MTT for cell metabolic activity, total amount of DNA, alkaline phosphatase activity and ACE activity. So far, we have used several cell lines.

While miniaturization is an improvement to the traditional approach of characterizing the bioresponse to polymer substrates, the ultimate in high-throughput characterization is the preparation of polymer film microarrays. Successful implementation of this technology requires that the polymer spots adhere strongly to the substrate, be stable under aqueous conditions of the bioassay, and be selectively reactive in the bioassay relative to the substrate to maximize the resolution of the analysis [41]. Specifically, microarrays of polyacrylates were prepared by synthesis of the spotted monomers at a density of 1728 spots on a 25 × 75 mm poly(hydroxyethyl methacrylate)-coated glass slide [41]. Cell growth and differentiation of human embryonic cells were studied simultaneously for the different biomaterials using the microarray. Cell attachment and spreading occurred on many of the polyacrylates, except those containing particular monomers. Alternatively, polymers can be transferred to generate microarrays, allowing for the facile preparation of thin films for the determination of cell-biomaterial interactions [43] or for screening for bacterial adhesion [44]. While this technology exists, the preparation of the polyacrylate microarrays revealed that the technique requires fine-tuning and that not all polymers were amenable to spotting uniformly [41].

7.5.2. Protein Adsorption

Protein adsorption to biomaterial surfaces is an important indicator of biocompatibility [1]. The adsorbed protein layer directs the adsorption affinity of other proteins and ultimately whether or not cells will attach to the material, a vital parameter for tissue scaffolds [45,46]. The adsorption of fibrinogen, a major protein component in blood serum, onto polyarylate surfaces was investigated since fibrinogen-coated surfaces are known to have decreased compatibility with blood [47]. A high-throughput method was developed to measure the adsorption of fibrinogen on biomaterial thin films using a fluorescence-based assay in a 384-well microtiter plate [42] since the standard techniques for measuring protein adsorption on artifical substrates, such as immunoblotting [48] or plasmon resonance [49], were labor-intensive and expensive. This miniaturized technique decreased the volume of assay reagents and utilized the standard microplate technologies, such as plate readers, automatic pipeting systems, and plate washers. The technique is not only amenable to automation, but can be further miniaturized to 1536-well and 3456-well formats [50]. Briefly, the level of fibrinogen adsorbed on the polymer films was determined by reaction with fluorescently-labeled (fibrinogen) antibodies. The level of fibrinogen adsorption was measured using a fluorescence plate reader, and the values were reported relative to the control, fibrinogen adsorption on naked polypropylene. The immunofluorescence assay proved very effective for determining whether fibrinogen adsorption was low or high on different polymer surfaces; however, smaller differences between polymers with similar fibrinogen adsorption could not be distinguished (Fig. 7.4). These differences were elegantly distinguished in a second, more involved assay using Q-sense [51]. While fibrinogen adsorption is thought to depend highly on surface hydrophobicity, a simple plot of air–water contact angle and fibrinogen adsorption revealed that the relationship cannot be predicted by a simple linear correlation. Interestingly, the diacids

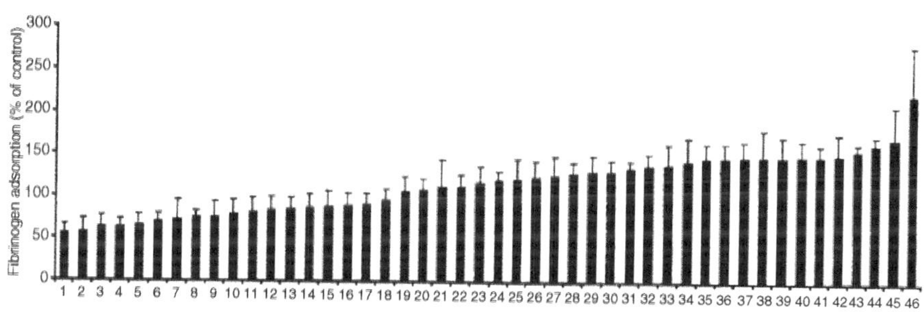

1. Poly(DTiB sebacate)
2. Poly(HTH sebacate)
3. Poly(DTO glutarate)
4. Poly(DTH adipate)
5. Poly(DTO sebacate)
6. Poly(HTH suberate)
7. Poly(DTO adipate)
8. Poly(DTO suberate)
9. Poly(HTH adipate)
10. Poly(DTBn sebacate)
11. Poly(DTH 3-methyladipate)
12. Poly(DTH suberate)
13. Poly(DTH glutarate)
14. Poly(DTiB adipate)
15. Poly(DTM sebacate)
16. Poly(DTsB suberate)
17. Poly(DTBn suberate)
18. Poly(HTF suberate)
19. Poly(DTsB adipate)
20. Poly(HTE adipate)
21. Poly(DTM 3-methyladipate)
22. Poly(DTH succinate)
23. Poly(DTB suberate)
24. Poly(DTiP adipate)
25. Poly(DTO succinate)
26. Poly(DTB succinate)
27. Poly(DTB glutarate)
28. Poly(DTM suberate)
29. Poly(DTBn adipate)
30. Poly(HTE 3-methyladipate)
31. Poly(DTE adipate)
32. Poly(DTM adipate)
33. Poly(DTsB 3-methyladipate)
34. Poly(DTO diglycolate)
35. Poly(DTsB 3-methyladipate)
36. Poly(DTBn 3-methyladipate)
37. Poly(DTiP 3-methyladipate)
38. Poly(DTB adipate)
39. Poly(DTB 3-methyladipate)
40. Poly(DTH diglycolate)
41. Poly(DTE 3-methyladipate)
42. Poly(DTE glutarate)
43. Poly(lactic glycolic acid)
44. Poly(lactic acid)
45. Poly(HTE succinate)
46. Poly(DTB diglycolate)

Figure 7.4. Adsorption of human fibrinogen on 44 polyarylates, PLA, and PLGA. The amount of adsorbed fibrinogen is material-dependent (mean ± SD; $n = 16$). The statistical comparison of each of 10 low fibrinogen-binding polymers (Nos. 1–10) with each of the high fibrinogen-binding polymers (Nos. 37–46) showed significant differences between the polymers ($p < 0.001$). (From Weber, N., Bolikal, D., Bourke, S. L., and Kohn, J., Small changes in the polymer structure influence the adsorption behavior of fibrinogen on polymer surfaces: Validation of a new rapid screening technique, *J. Biomed. Mater. Res.* **68A**:496–503. Copyright © 2004 Wiley Periodicals, Inc.)

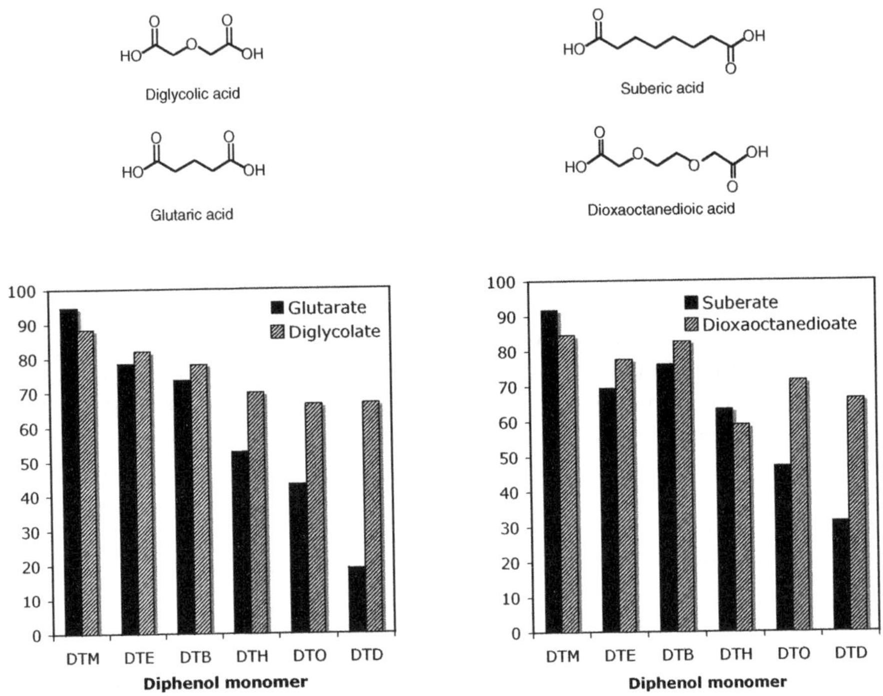

Figure 7.5. Normalized fibroblast proliferation on two polyarylates differing by only one atom in the diacid: poly(DTR glutarate) vs. poly(DTR diglycolate) and poly(DTR suberate) vs. poly(DTR dioxaoctanedioate).

containing an oxygen atom in their backbone adsorbed the most fibrinogen, despite their overall measure of hydrophobicity (Fig. 7.5). The authors hypothesize that the oxygen affects the mode of fibrinogen binding. Computational modeling of these data is discussed later in this chapter.

7.5.3. Cell Proliferation

Biocompatibility can also be measured using cell response studies. The response of fetal rat lung fibroblasts (FRLF) (RLF-1) to polymeric substrates was determined using an MTS colorimetric assay (Promega, Madison, WI) [17]. This assay detected whether the biomaterial was cytotoxic to fibroblast cells or whether it supported proliferation of the cells. Quantitative results for the cellular response to each polyarylate sample were obtained by reporting the *normalized metabolic activity* (NMA), which was defined as the average measured metabolic activity relative to the average value determined for the control, tissue culture polystyrene (TCPS). Likewise, the metabolic activity of a modified line of rat lung fibroblasts (RLF) and normal foreskin fibroblasts (NFF) on polyarylate surfaces was determined. The experimental results were

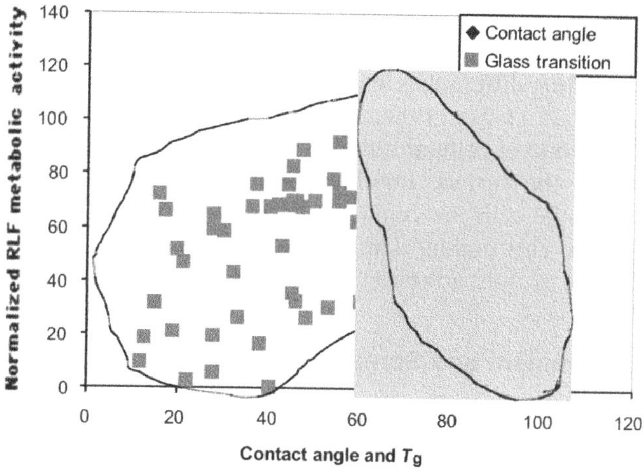

Figure 7.6. The shaded regions represent the total possible values for glass transition temperature and contact angle in the family of polyarylates. The region where they overlap describes the polymer library space (and the property–performance correlation) where cell proliferation is favored.

used to develop a semiempirical computational model, which was able to predict biomaterials that would support RLF cell growth [52]. High RLF metabolic activity was observed when the polymer was made up of either DTM, DTE, or HTE, but not adipic acid or 3-methyl adipic acid, and the air–water contact angle was less than or equal to 69, the glass transition temperature was greater than 65, and TFI was less than or equal to 7. These results are summarized in Figure 7.6, which shows the region in polymer space for overlap of the glass transition temperature and the air–water contact angle versus metabolic activity. This assay was more recently adapted from a 24-well plate to a 96-well plate format to maximize efficiency [19]. Specifically, the statistical significance of cellular response was improved by increasing the number of repeats nearly threefold, while simultaneously reducing the screening from 3 months to 1 week and decreasing the cost per test by over 600%. Both validation of the cell lines and application of the Ryan–Joiner test for Gaussian distribution of cell growth were necessary to obtain reproducible results that were amenable to standard statistical analysis [53].

7.5.4. Surface Roughness

The development of methods to isolate certain material phenomenon, such as surface roughness, is essential to effectively probe the biological response to structural variations. The effect of surface roughness on proliferation of MC3T3-E1 osteoblastic cells was determined in real time using a substrate with a gradient of surface roughness, ranging from roof mean square (rms) = 0.5–13 nm) [54,55]. The samples were prepared by gradient annealing of a polymer film, which affects the size of the crystallites and the surface roughness in the *x* direction. The ability to perform the combinatorial bioassay improved reproducibility by avoiding differences in cell population and culture conditions. The results revealed that it was not the concentration or conformation of adherent serum proteins, mode of cellular adhesion, cell shape, or rearrangement of the cytoskeleton, but the surface topology that affected cell proliferation. The results show that the cells are remarkably sensitive to surface topography on the order of 5 nm. This finding demonstrates processing of a biomaterial can be used to control protein adhesion on implantable devices.

7.5.5. Cell Attachment and Spreading

Cell attachment and spreading assays have been developed for a 96-well polypropylene microplate [45]. The characterization of tissue spreading on a substrate is vital to the development of biomaterial scaffolds for tissue engineering or artificial organs. The rate of tissue spreading depends on both the cell–substrate adhesivity and the cell–cell cohesivity. The overlapping effect was decoupled by using a set of structurally related poly(DTE-co-PEG carbonate)s (PEG = 1000 g/mol, 0–5 mol%) as substrates that would have different adhesivity to cells and by using cell lines that exhibited different levels of cell–cell cohesivity. Tissue surface tensiometry was used to measure the cell–cell cohesivity, and the cell–substrate adhesivity was determined by measuring the number of cells attached following inversion of the culture plate after different incubation times using a fluorescence microplate reader. Cell spreading with low cell–cell cohesivity decreased with increasing PEG content using confocal microscopy. Furthermore, cells with high cell–cell cohesivity did not spread on the most adhesive substrate, typifying the importance of understanding cell spreading in designing substrates for tissue engineering. Choosing cell lines with cell–cell cohesivity similar to that desired for a tissue engineering application may provide a method to rapidly screen a material for time-efficient tissue spreading.

7.5.6. Gene Expression

The presence of an inflammatory response directly determines the suitability of material for use as an implantable device. Since inflammation is induced by macrophages, a type of white blood cell, it is important to understand the interaction of macrophages with biomaterials surfaces. Specifically, the interaction of macrophages with a foreign body results in a cascade of events, which can lead to the release of proinflammatory or antiinflammatory cytokines. In order to design biomaterials with minimum rejection, it is important to be able to screen polymers for their ability to induce the production of proinflammatory cytokines, such as IL-1β or IL-6, in the macrophage. Gene expression of these two cytokines was determined relative to β-2-microglobulin on polyarylate surfaces using real-time reverse transcriptase-polymerase chain reaction (RT-PCR) [19]. Briefly, reverse transcriptase is an enzyme that utilizes an

RNA template to synthesize a complementary strand of DNA (cDNA), which is amenable to accurate, million-fold amplification of the complementary strand (DNA version of the original RNA) in hours using specific oligonucleotide primers with DNA polymerase in a process called *polymerase chain reaction* (PCR) [56]. The amplification incorporates fluorescent labels in the amplified cDNA for quantification in real time by in situ fluorescence detection. Measurements made in real time are more reliable than endpoint measurements of PCR products using traditional RT-PCR. There are many advantages to quantifying gene sequences using this technology such as sensitivity and precision, fast turnaround time for data acquisition and analysis, and reduction of labor time and costs [57,58].

RT-PCR was performed on macrophages that were cultured with and without lipopolysaccharide. The former provide a basal level for the cytokines, and the latter induced the activation of the macrophages, resulting in the release of growth factors that encourage synthesis of the extracellular matrix [59]. Both cytokines, IL-1β and IL-6, were upregulated on nearly all polymer substrates in the assays with activated macrophages (compared to basal levels) (Fig. 7.7). Interestingly, only poly(DTBn dodecanoate) inhibited IL-1β

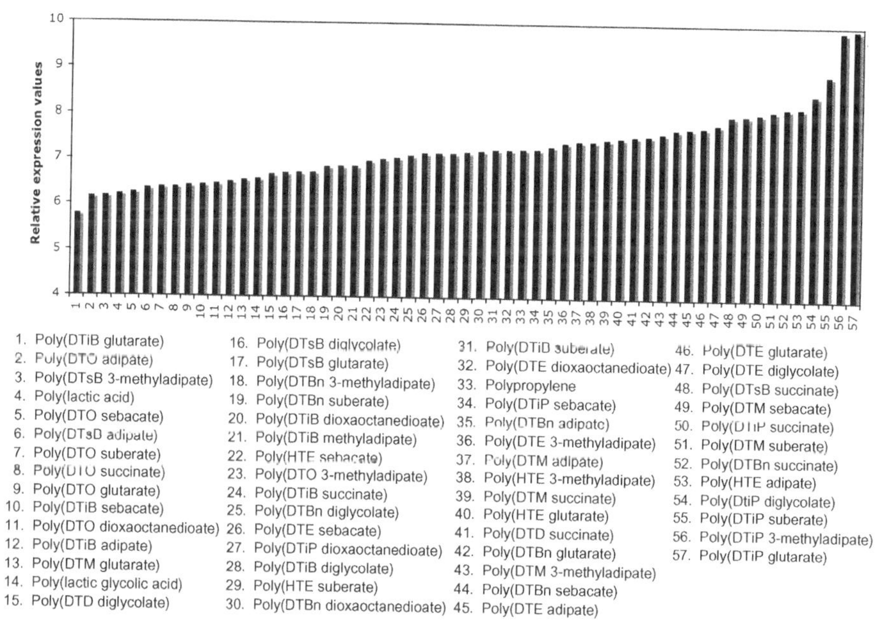

Figure 7.7. Real-time RT-PCR results for gene expression of IL-1β cytokines on different polyarylate substrates. The black bars represent the basal levels of these genes in the quiscient state [without lipopolysaccharide (LPS)]. [Reproduced with permission from Smith, J. R., Seyda, A., Weber, N., Knight, D., Abramson, S., and Kohn, J., Integration of combinatorial synthesis, rapid screening, and computational modeling in biomaterials development, *Macromol. Rapid Commun.* **25**:127–140. Copyright © 2004 Wiley-VCH Verlag GmbH & Co KG.]

expression. While the general trend of upregulation for these two proinflammatory cytokines was consistent, no correlation between structure and IL-6 expression was determined. A more complex correlation of polymer structure and the level of gene expression of these two cytokines in macrophages on different polymer surfaces can potentially be related to other experimental values, such as protein adsorption, cell proliferation, and glass transition temperature using computational models.

The advances in high-throughput characterization have been substantial since the mid-1990s. Recent developments in combinatorial materials science have been the focus of several recent articles and was the dedicated focus of a special issue of *Measurement Science and Technology* [15,60,61]. Examples of high-throughput characterization techniques available include Rapid GPC™ systems, which screen molecular weight in 40 s–2 min; MALDI-TOF-MS; and optical screening techniques, including ATR-FTIR, photoluminescence, and UV-visible spectroscopy; in addition to atomic force microscopy and optical microscopy [60]. In general, these techniques fall into three categories: parallel, semiautomated, and automated approaches. The move to completely automated approaches whereby samples can be taken from the reaction mixture and prepared for testing, with data logged automatically, and data presented in an organized graphical format without any manual intervention, will be the point where high throughput characterization can truly support combinatorial synthesis of materials [62].

7.5.7. Practical Application of the Combinatorial Approach to Identify "Lead" Polymers

One of the basic premises of "CombiChem" is the utilization of a rapid assay that can identify lead compounds for further study. In the case of polymers for use in tissue regeneration scaffolds, one of the important requirements of the scaffold in vivo is to assist in the reorganization of the initial cell population into functional (i.e., differentiated) tissue. Thus, on an intuitive level, one may be tempted to formulate the hypothesis that those polymeric surfaces that supported early differentiation in vitro may be promising candidates for investigation as potential scaffold materials in vivo. This rationale was tested in an attempt to identify osteoconductive polymers among the library of polyarylates.

An initial evaluation of the rate of proliferation of an osteoblast-like cell line (UMR106) grown on compression molded, flat films of various polyarylates determined the polymers supported cell proliferation with no statistically significant variations between the test polymers [63]. Next, alkaline phosphatase activity, an early marker of differentiation for osteoblast-like cells, was determined to evaluate the ability of test surfaces to induce cell differentiation [63]. Cells grown on eight specific polymer surfaces showed significantly higher levels of alkaline phosphatase activity compared to the other test polymers or tissue culture polystyrene. The most striking feature common to five of the eight potential lead polymers was that nearly all of the polyarylates contained DTE monomer. In order to screen these polymers efficiently in vivo for osteoconductivity, a high-throughput screening implantation device was developed by modifying the bone cage model [64]. Briefly, a polyethylene chamber was divided into 10 thin, narrow channels, each lined with a different type of polymer using two thin coupons of each test material. Since one chamber can be implanted into each femur, 20 polymers can be tested in one animal. Once the chamber is implanted into the femur of a canine (Fig. 7.8), bone can grow into the channels from the top and the bottom. The total amount of bone ingrowth into the channels, measured by acoustic scanning microscopy after explantation, reflects the bone conductivity and hard tissue biocompatibility of the test material (Figs. 7.9 and 7.10) [65,66]. The DTE-containing polymers exhibited superior osteoconductivity in this in vivo screening, suggesting that combinatorial approaches can benefit the discovery of new biomaterials.

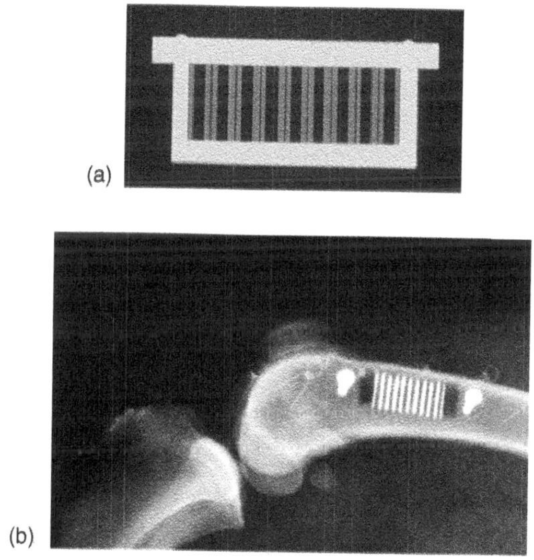

Figure 7.8. (a) The Alexander cage model; (b) the Alexander cage in place. Previously, this model had been used exclusively in such a way that all coupons used were of the same test material, producing 10 identical channels. This model was adapted as a screening assay for osteoconductivity by using coupons made of different materials so that each channel was lined by a different test material. In this way, up to 20 different materials can be screened for osteoconductivity in vivo using one single animal. [Images provided courtesy of Professor Russell Parsons, New Jersey Medical School, University of Medicine and Dentistry of New Jersey (UMDNJ).]

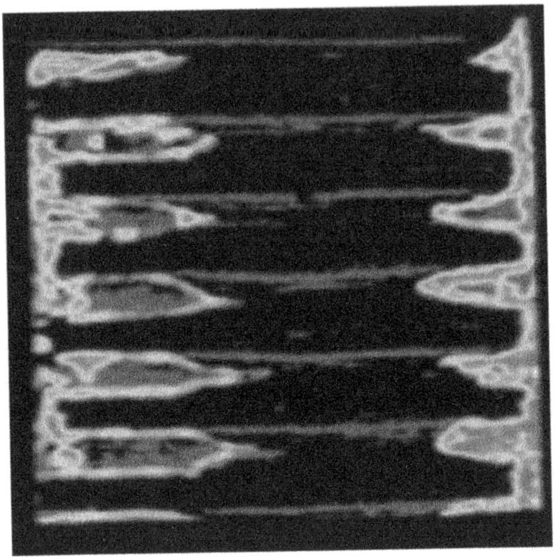

Figure 7.9. Acoustic scanning microscopy is based on the fact that the strength of the reflected acoustic signal is related to the stiffness of the reflecting material. Thus, soft tissue appears black and an increase in stiffness is indicated by an artificial color scale. Orange and red represent fully mineralized bone. The chamber shown here contains identical coupons of poly(L-lactic acid), a marginally osteoconductive material. The model is highly reproductive as shown by the virtually identical levels of bone ingrowth into each of the channels. [Image provided courtesy of Professor Russell Parsons, New Jersey Medical School, University of Medicine and Dentistry of New Jersey (UMDNJ).]

7.6. COMPUTATIONAL MODELING

7.6.1. Overview

The complexity of biological behavior precludes the development of strict ab initio models from which the entire set of bioresponses to biomaterials can be predicted without recourse to experimental input. Consequently, it is necessary to develop *surrogate models* [also known as *quantitative structure–activity relationships* (QSARs) or *quantitative structure–performance relationships* (QSPRs)], which combine experiment and theory to achieve an accurate representation of bioresponse to biomaterials. This approach has been effectively used in virtually all engineering and science disciplines.

The Surrogate Model concept is illustrated in Figure 7.11. A specific measurable bioresponse (e.g., mass of adsorbed fibrinogen according to a defined laboratory protocol) is selected and a polymer library is identified (e.g., polyarylates). It is assumed that the quantitative bioresponse for an individual polymer is determined by the polymer's chemical structure (for the defined laboratory protocol). Therefore, in principle, a quantitative model can be developed relating the bioresponse to the known chemical structure that is

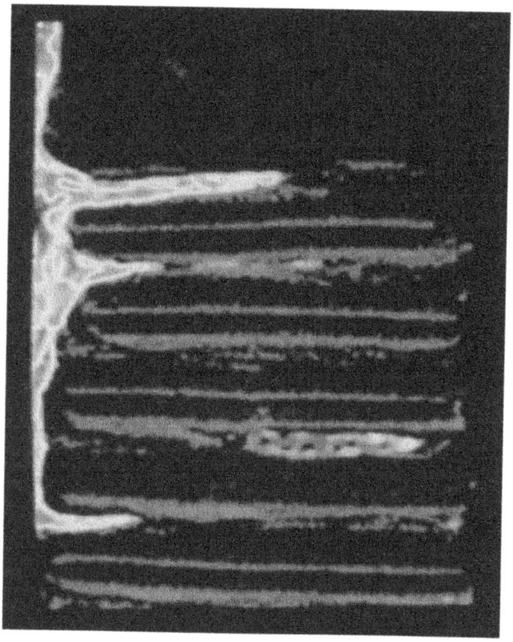

Figure 7.10. Qualitative screening for osteoconductivity using a modification of Alexander's bone chamber model. Here, coupons made of polymers with different pendent chains were used to generate channels that were lined by different test materials. The channel with the highest degree of bone ingrowth was lined with coupons made of a polymer containing the "DTE" unit. [Image provided courtesy of Professor Russell Parsons, New Jersey Medical School, University of Medicine and Dentistry of New Jersey (UMDNJ).]

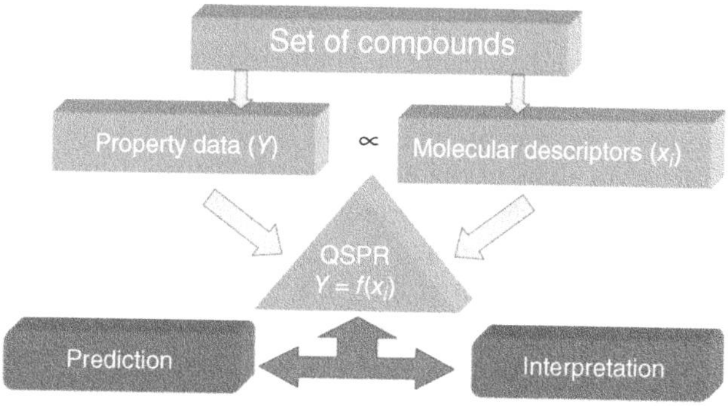

Figure 7.11. Quantitative structure–property relationship (QSPR).

described in terms of *chemical descriptors*. A wide range of descriptors have been developed (see, e.g., Ref. 67). Examples include molecular density, logarithm of the octanol/water partition coefficient (logP), and the number of single bonds. Commercial software (e.g., Molecular Operating Environment© [68] and Dragon© [69] generate hundreds of descriptors for a given polymer chemical structure. A surrogate model is developed to predict the specific bioresponse for a given polymer in terms of its chemical descriptors. A variety of different mathematical forms may be used for the surrogate model, including an *artificial neural network* (ANN) [70], *support vector machine* [71], and *partial least squares* [72]. The model is *trained* using a selected subset of experimental data that (ideally) represents the entire range of bioresponse. The model is *validated* by comparison of the predicted and experimental results for the remainder of the experimental dataset.

A surrogate model has two principal advantages; it can (1) predict a specific bioresponse for any polymer within a library on the basis of experimental data for a subset of the polymers (this is particularly important for large libraries for which an exhaustive experimental survey would be costly and time-consuming) and (2) identify the principal features of the polymer chemistry that most strongly affect the specific bioresponse, and lead to further insights into polymer design for specific biomedical applications.

7.6.2. Surrogate Models

Two commonly used mathematical forms for surrogate models are artificial neural networks and partial least squares. We present the basic mathematics for each of these models and refer the reader to the references cited herein for further details.

7.6.2.1. Artificial Neural Network. An artificial neural network (ANN) is ideally suited to serve as a surrogate model since it is capable of representing a complex, nonlinear behavior to an arbitrary level of accuracy [70]. An ANN is a nonlinear mathematical relationship between input variables (i.e., chemical descriptors) and the output variable (i.e., the specific bioresponse). The ANN incorporates *weights* that are determined by minimizing the difference between the predicted and experimental bioresponse for a subset of the experimental data (the *training set*). The quality of the ANN is determined by the average error between the predicted and experimental values of the bioresponse for the remainder of the experimental dataset (the *validation set*).

The structure of an ANN is shown in Figure 7.12. The input is a vector x of length $m + 1$ consisting of m descriptors plus a constant offset. The input vector defines the specific polymer in terms of m chemical descriptors, where the number m and type of descriptors may be determined by a separate algorithm (see below). The output is a scalar prediction, z, of the bioresponse for the particular polymer represented by the descriptors. The vector of input variables is denoted as follows:

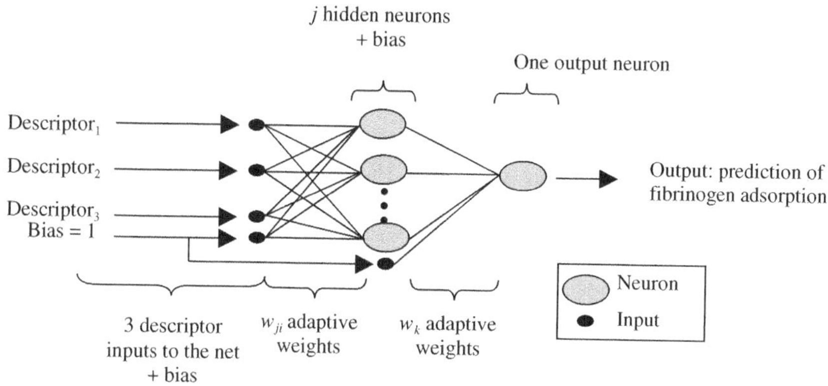

Figure 7.12. Schematic of ANN.

$$x = (x_0, x_1, \ldots, x_i, \ldots, x_m) \tag{7.1}$$

The additional term x_0 is included to provide a constant offset and assigned the value $x_0 = 1$. The vector of input variables for the kth polymer, x_k, is denoted by

$$x_k = (x_{k,0}, x_{k,1}, \ldots, x_{k,i}, \ldots, x_{k,m}) \tag{7.2}$$

The vector of hidden variables, y, is denoted by

$$y = (y_0, y_1, \ldots, y_i, \ldots, y_m) \tag{7.3}$$

where n is the number of hidden variables. Typically, n is 2 or 3, and the accuracy of the ANN is usually found to be insensitive to the value of n. The additional term, y_0, is included to provide a constant offset and assigned the value $y_0 = 1$. The vector of hidden variables, y_k, for the kth polymer is denoted by

$$y_k = (y_{k,0}, y_{k,1}, \ldots, y_{k,i}, \ldots, y_{k,m}) \tag{7.4}$$

The hidden variables y_k for the kth polymer are calculated according to

$$y_{k,j} = f\left(\sum_{i=0}^{m} w_{j,i}^{0} x_{k,i}\right) \quad \text{for} \quad k = 0, \ldots, s-1 \quad \text{and} \quad j = 1, n \tag{7.5}$$

where s is the number of polymers and the vector of weights $w^0{}_j$ for the jth hidden layer neuron is:

$$w_j^0 = (w_{j,0}^0, w_{j,1}^0, \ldots w_{j,i}^0, \ldots, w_{j,m}^0) \quad for \quad j = 1, \ldots, n \tag{7.6}$$

The function f, defined by

$$f(\xi) = \left(1 + e^{-\kappa\xi}\right)^{-1} \tag{7.7}$$

varies between 0 and 1 and is known as the *sigmoid function*. The quantity κ is user-specified. The output value for the kth polymer is denoted z_k and defined by

$$z_k = \sum_{j=0}^{n} w_j^l y_{k,j} \tag{7.8}$$

where the vector of weights w^1 for the output layers is

$$\vec{w}^1 = \left(w_0^1, w_1^1, \ldots, w_j^1, \ldots w_n^1\right) \tag{7.9}$$

There are a total of $n(m + 2) + 1$ unknowns in the ANN, namely, $n(m + 1)$ values of w^0 and $n + 1$ of w^1. The weights w^0 and w^1 are obtained by minimizing the square error E defined by

$$E = \frac{1}{2}\sum_{k=0}^{s-1}(z_k - \hat{z}_k)^2 \tag{7.10}$$

for a fraction of the experimental dataset (e.g., one-half of the polymers selected at random). This set is identified as the *training set*, and the remaining polymers are denoted the *validation set*. In Eq. (7.10), the experimental value for the kth polymer is denoted by $\hat{z}_k$ and the predicted value by z_k. The minimization can be achieved using a variety of optimization algorithms including *backpropagation* [70] or a *genetic algorithm* [73].

The accuracy of the ANN is evaluated by comparison of the predicted and experimental bioresponse for the remaining fraction of the polymers (the *validation set*). Two measures of accuracy are used. The first measure is the percent rms relative error $E_{\rm rms}$ defined by

$$E_{\rm rms} = 100 \times \sqrt{\frac{1}{s_{\rm v}}\sum_{k=0}^{s-1}\left(\frac{z_k - \hat{z}_k}{\hat{z}_k}\right)^2} \tag{7.11}$$

where $s_{\rm v}$ is the number of polymers in the validation set. The second measure is the Pearson correlation coefficient:

$$\rho = \frac{1}{\sigma\hat{\sigma}}\sum_{k=0}^{x_{\rm v}-1}(z_k - \bar{z}_k)\left(\hat{z}_k - \bar{\hat{z}}\right) \tag{7.12}$$

where $\bar{z}$ and $\bar{\hat{z}}$ are the mean predicted and experimental values for the validation set, and σ and $\hat{\sigma}$ are the respective standard deviations:

$$\sigma = \sqrt{\frac{1}{s_v}\sum_{k=0}^{s_v-1}(z_k - \bar{z})^2} \tag{7.13}$$

There are three user-specified parameters in the ANN: (1) the number of descriptors m, (2) the number of neurons n in the hidden layer, and (3) the inverse length scale κ in the sigmoid function. An ANN is useful provided that the predicted results are insensitive to the choice of these parameters within reasonable ranges.

The selection of the descriptors is an important step in the development of the ANN. Commercial software (e.g., Molecular Operating Environment© and Dragon©) generate hundreds of descriptors. It is intuitively unlikely, however, that the descriptors are of equal significance in the model, and furthermore it is well known that the use of an excessive number of input variables leads to overfitting of the experimental data and a reduction in accuracy of the ANN as evaluated using the validation set. Therefore, the number of input variables must be reduced from hundreds to the order of ten or less. Two methods have been used: (1) decision tree and (2) partial least-squares regression.

The C5 decision tree routine [74] calculates the information gain of each of the descriptors with respect to the specific bioresponse. Information gain (IG) is a measure of the ability of a particular attribute to classify a sample set [75], specifically, the decrease of the weighted average disorder of classification after the set has been partitioned by that descriptor. Each of the polymers is classified based on the value of its specific bioresponse relative to fixed set of subranges of the experimental values. For example, if the overall range of experimental values is divided into five "bins" of equal size, then each polymer is classified based on the bin corresponding to its specific bioresponse value. The bin size (e.g., 20%) is chosen to be approximately the same as the experimental uncertainty. Then, the IG of particular descriptor (D) for the entire experimental dataset (E) is calculated according to

$$\mathrm{IG}(E,D) = \mathrm{entropy}(E) - \sum_{i=1,\ldots,n}\left(\frac{|E_i|}{|E|}\right)\mathrm{entropy}(E_i) \tag{7.14}$$

where n = number of (user-defined) different partitions of descriptor in sample set
E_i = set of samples that have a descriptor value within partition i
$|X|$ = number of samples in set X

and

$$\mathrm{Entropy}(E) = -\sum_{j=1,\ldots,5}\left(\frac{|E_j|}{|E|}\right)\log_2\left(\frac{|E_j|}{|E|}\right) \tag{7.15}$$

The sum in (7.14) is over all descriptor values in the sample set, while the summation in (7.15) is over all five classes ("bins") in the experimental dataset. The descriptors are ranked in terms of their IG value, and the highest-ranking descriptors are selected as input variables for the ANN. In addition, sensitivity studies can be performed to ascertain the effect of increasing the number of descriptors from, say, 3 to 10.

7.6.2.2. Partial Least Squares. Partial least squares (PLS) is a linear multivariate regression method that can be used either as a surrogate model or as another method for selecting the appropriate subset of descriptors as input variables for an ANN. PLS was developed by Wold [76]. PLS develops a linear expression for the specific bioresponse z in terms of linear combinations (known as *principal components* [77]) of the descriptors

$$z = a_0 + a_1(\mathrm{PC}_1) + a_2(\mathrm{PC}_2) + \cdots + a_n(\mathrm{PC}_n) \tag{7.16}$$

where n is the number of descriptors. PLS orders the contribution from successive terms in decreasing absolute magnitude, and therefore the first few terms (e.g., PC_1 and PC_2) identify the descriptors that have the greatest influence on the predicted bioresponse. The PLS surrogate model can then be truncated to the first few terms in Eq. (7.16). Alternatively, the descriptors contained within the first (or, if desired, first several) principal components can then be selected as input variables for the ANN, and the ANN is trained in the usual manner.

Partial least squares is based on *multiple linear regression* [78]. The output z is assumed to be a linear function of the input variables (descriptors) according to

$$z = b_0 + b_1 x_1 + b_2 x_2 + \cdots + b_m x_m \tag{7.17}$$

This differs from the ANN formulation wherein the output z is a *nonlinear* function of the input variables as indicated by Eqs. (7.5) and (7.8). The regression coefficients b_j for multiple linear regression as follows. Consider n measurements z_k for $k = 1, \ldots, n$. The corresponding values of the input variables are denoted by the subscript 2. The error between the predicted and measured output is denoted by e_k. Thus

$$z_k = b_0 + b_1 x_{k,1} + b_2 x_{k,2} + \cdots + b_m x_{k,m} + e_k \tag{7.18}$$

Denoting the set of all n measurements as a column vector

$$Z = \begin{Bmatrix} z_1 \\ \vdots \\ z_n \end{Bmatrix} \tag{7.19}$$

the set of n input variables as a matrix

$$X = \begin{Bmatrix} 1 & x_{1,1} \cdots x_{1,m-1} & x_{1,m} \\ 1 & x_{2,1} \cdots x_{2,m-1} & x_{2,m} \\ & \ddots & \\ 1 & x_{n,1} \cdots x_{n,m-1} & x_{n,m} \end{Bmatrix} \tag{7.20}$$

the set of errors as

$$E = \begin{Bmatrix} e_0 \\ e_1 \\ \vdots \\ e_m \end{Bmatrix} \tag{7.21}$$

and the set of $m + 1$ regression coefficients as

$$B = \begin{Bmatrix} b_0 \\ b_1 \\ \vdots \\ b_m \end{Bmatrix} \tag{7.22}$$

then the rms error is minimized by choosing

$$B = (X'X)^{-1} X'Z \tag{7.23}$$

where X' is the transpose of X [provided the inverse $(X'X)^{-1}$ exists]. Partial least squares then incorporates the possible correlations between input variables by computing a *factor score matrix* T defined by

$$T = XW \tag{7.24}$$

where W is an appropriate weight matrix, and then generates the linear regression model

$$Z = TA + E \tag{7.25}$$

where

$$B = WA \tag{7.26}$$

The factor score matrix T represents the principal components of the descriptors. The factor score matrix is computed using the *nonlinear iterative partial least-squares* algorithm [78].

7.6.3. Results

7.6.3.1. Prediction of Fibrinogen Adsorption. An ANN surrogate model was developed for prediction of fibrinogen adsorption for a library of tyrosine-derived polyarylates. Fibrinogen adsorption was measured for 45 polymers using an microtiterplate-based immunofluorescence assay [42]. The average uncertainty in the experimental measurement was 17.9%, and the range of adsorption was 366%. A total of 106 descriptors were generated for each of the 45 polymers, with 104 descriptors computed using the commercial software Molecular Operating Environment© based on the polymer chemical structure viewed as a linear polymer (i.e., without consideration for three-dimensional conformational effects) and two descriptors obtained experimentally (specifically, the glass transition temperature T_g and air-water contact angle θ). The top three descriptors identified by the decision tree are T_g, the number of hydrogen atoms in the monomer (a_nH), and the logarithm of the octanol/water partition coefficient [79]. One-half of the library of 45 polymers was selected at random to train the ANN, and the accuracy of the ANN was evaluated by comparison of the predicted and experimental data for the remaining half. The results are shown in Figure 7.13. The ANN predicts the fibrinogen adsorption for 70% of the polymers in the validation set within the experimental uncertainty. The rms relative error is 38%, which compares favorably with the experimental uncertainty (17.9%) and is an order of magnitude smaller than the range of data (366%). The Pearson correlation coefficient is 0.54.

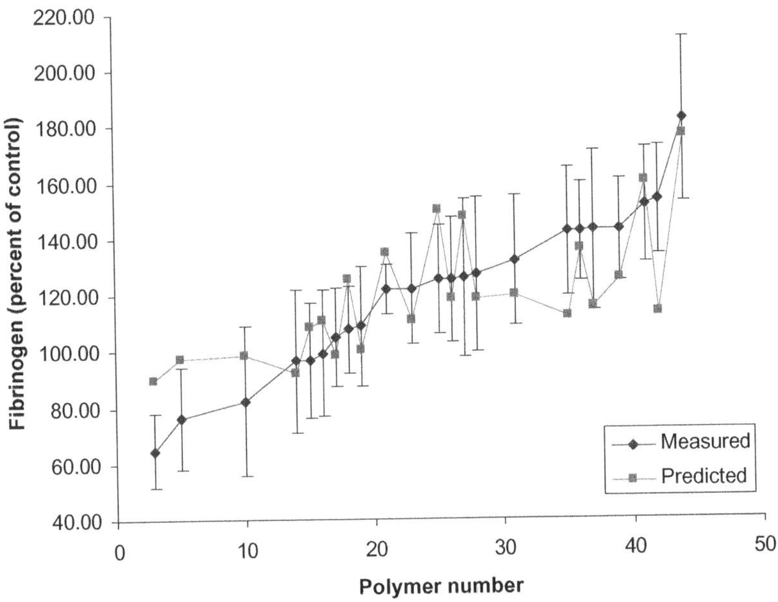

Figure 7.13. Validation set for ANN.

One effective way to spot trends in the structure–performance relationship model is to inspect the structures and associated descriptors for the extreme performers, say, the top five and bottom five polyarylates in terms of fibrinogen adsorption (Table 7.2). A conspicuous difference is seen in the pendant R group: fibrinogen adsorption is much higher for polyarylates having short R groups (average = 170.10%) than for those having longer, more extended R groups (average = 64.24%). Polymers with shorter R groups would be expected to pack more efficiently by virtue of their more cylindrical surface, thus enabling the polymer to form form strong van der Waals and hydrogen-bonding interactions with fibrinogen. Shorter R groups will also reduce the overall hydrophobicity of the polymer. Likewise, the nature of the diacid (Y) group shows clear trends among the top five and bottom five performers. Shorter and, therefore, less hydrophobic diacid groups exhibit better fibrinogen adsorption. The most striking example is DTB diglycolate, containing the most hydrophilic diacid group [Y = —C(O)—CH_2—O—CH_2—C(O)—] in this series of polyarylates, which elicited the highest measured fibrinogen adsorption by a considerable margin over the next-best polymer HTE succinate. Taken together, short R groups and short diacid (Y) groups will reduce polymer hydrophobicity due to the greater proportion of oxygen atoms in the repeat unit and will lead to a more compact, less flexible polymer. These trends are reflected in the corresponding values of the descriptors T_g, AWCA (θ), a_nH, and logP(o/w). Indeed, comparison of the averages for each descriptor between the top five and bottom five performers is instructive. Enhanced fibrinogen adsorption is associated with higher values of T_g and lower values of AWCA, a_nH, and logP(o/w). Higher T_g values are typically indicative of a conformationally less flexible or stiffer polymer chain and, all other things being equal, more closely packed chains conducive to strong interchain interactions. Correspondingly, those polyarylates with shorter diacids (Y) and less extended pendant groups (R) are expected to possess less conformational flexibility and more efficient chain packing. The lower values of the AWCA (θ) and logP(o/w) for the top five polyarylates denote reduced hydrophobic character, again consistent with shorter diacids and with shorter pendant groups that contain a higher proportion of O atoms. Subsequent analysis indicated that AWCA (θ) and logP(o/w) encoded similar information about polymer hydrophobicity and, hence, θ can be removed as a descriptor from the surrogate model without appreciable loss in predictability. The elimination of θ as an independent variable is significant, as it removes the need for a measured parameter. Indeed, the ultimate goal is to generate a surrogate model in which all of the independent variables are derived computationally. The lower values of a_nH for the top performers are indicative of polyarylates with shorter, more compact pendent groups (R) and diacid backbones (Y). Taken together, these results provide useful clues for the rational design of polymers whose structural features lend themselves to optimal biological and materials properties.

TABLE 7.2. Influence of Length of Diacid Component (Y) and Pendant Chain (R) on Adsorption of Fibrinogen[a]

Diacid Component (Y)	Pendant Chain (R)**	Fibrinogen Adsorption (%)	T_g (°C)	AWCA (°)	a_nH	logP(o/w)
Top 5 Polyarylates in Terms of Fibrinogen Absorption						
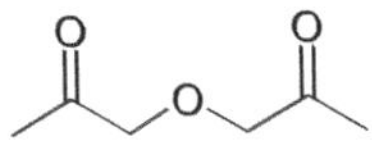	 C_4H_9	206.91	64	72	29	3.30
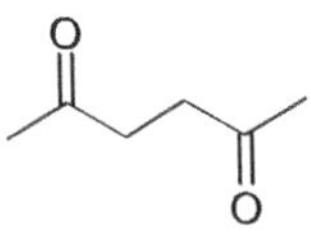	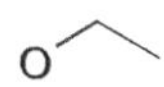 C_2H_5	182.15	73	68	23	2.90
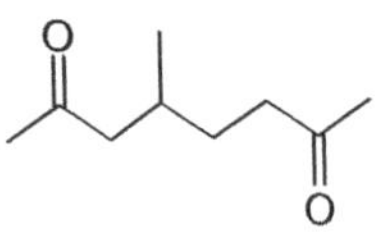	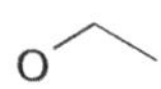 C_2H_5	156.74	63	75	31	4.23
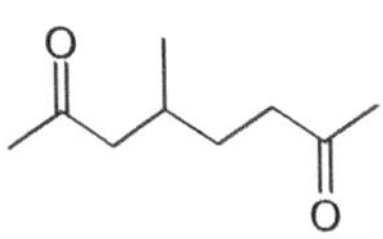	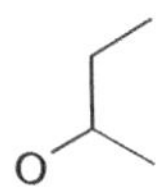 C_4H_9	153.27	56	76	35	5.31
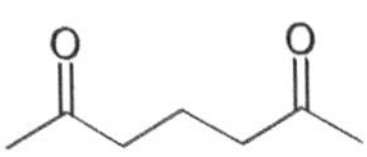	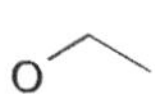 C_2H_5	151.44	69	69	27	3.43
Average		170.10	65	72	29	3.83
Bottom 5 Polyarylates in Terms of Fibrinogen Absorption						
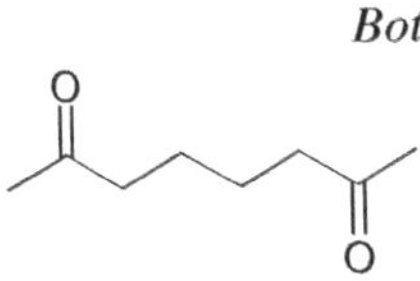	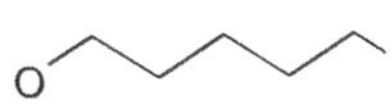 C_6H_{13}	76.20	32	84	35	5.72
	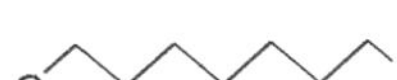 C_6H_{13}	66.00	13	96	49	8.46
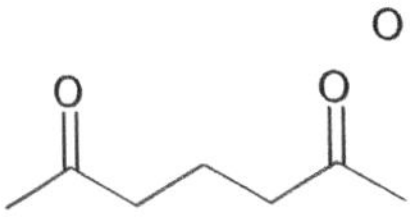	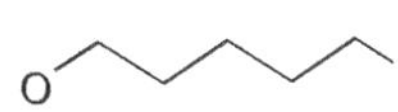 C_6H_{13}	64.80	32	86	39	6.25
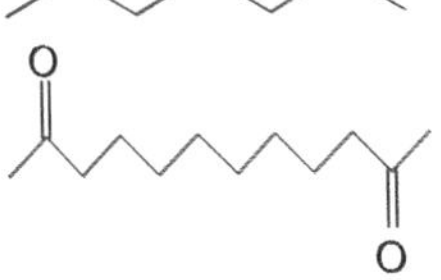	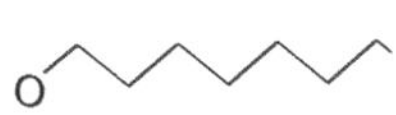 C_6H_{13}	57.60	23	86	43	7.49
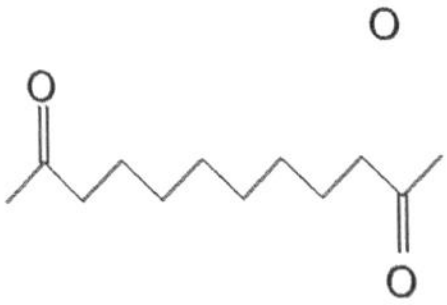	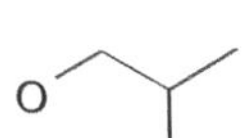 C_6H_{13}	56.6	33	83	41	6.61
Average		64.24	27	87	41	6.91

[a] LogP(o/w) and a_nH values given per monomer.

7.6.3.2. Prediction of Cell Proliferation. A PLS surrogate model was developed for prediction of cellular proliferation for a library of tyrosine-derived polyarylates [80]. Fetal rat lung fibroblast (FRLF) proliferation was measured for 62 polymers using a commercially available MTS colormetric assay [18]. The average uncertainty in the experimental measurement was 23%, and the range of proliferation was 5700%. A total of 11 descriptors were generated for each of the 62 polymers using the commercial software Dragon©. The optimal subset of descriptors, determined by maximizing the cross-validated correlation coefficient, comprised the number of tertiary carbons, number of branches in the pendent chain (aliphatic), molar refractivity, polar surface area, and logarithm of the octanol/water partition coefficient (logP). The predicted normalized metabolic activity [the ratio of cellular proliferation for a given polymer to the value for tissue culture polystyrene (TCPS)] is shown in Figure 7.14. The cross-validated correlation coefficient $q^2 = 0.56$ satisfied the generally accepted criterion that $q^2 \geq 0.50$ for an internally consistent and predictive regression model.

The PLS surrogate model generated predicted NMA values for the remaining 50 polymers in the library, and six representative and previously untested polymers were selected for subsequent experimental valuation. Two polymers each were selected from the lowest, middle, and highest ranges of NMA. The average percent error in the prediction is 15.8%, which is considerably below the experimental uncertainty (23%). The results, shown in Figure 7.15, clearly indicate the value of surrogate models.

Regarding the significance of the molecular descriptors employed to build the surrogate model, we note that descriptors closely related to polar surface area (PSA), molar refractivity (MR), and bioavailability (logP) were found in a previous study [81] to figure prominently in QSAR models constructed to

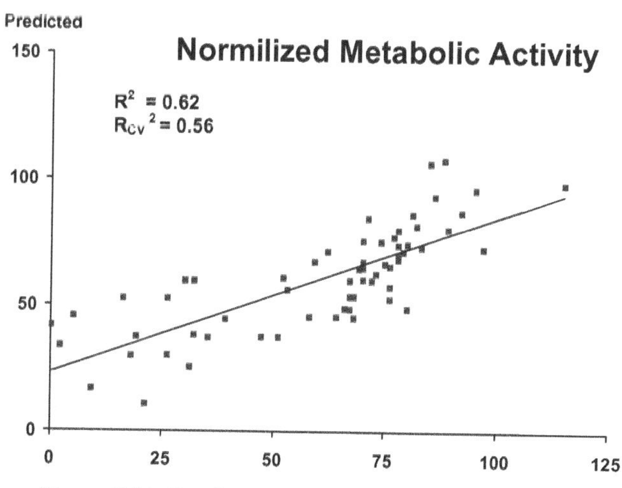

Figure 7.14. Predicted normalized metabolic activity.

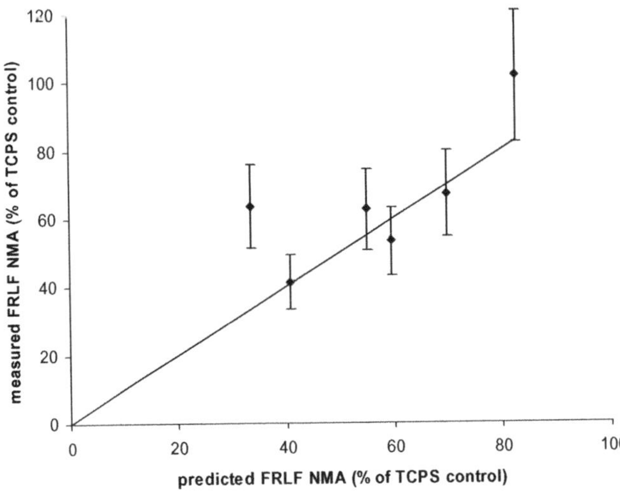

Figure 7.15. PLS surrogate model predictions for six previously untested polymers.

explain the observed inhibitor–receptor binding affinities for a series of HIV-1 protease inhibitors. We believe that the similarity of descriptors speaks to the fundamental importance of such computed structure-based descriptors in explaining the in vitro properties and, perhaps, the in vivo properties of molecules intended for biomedical applications.

The surrogate model can be used to distinguish the highest performers from the lowest. In other words, this model can be used to rapidly predict the performance of candidate structures prior to chemical synthesis. The predictions generated by the surrogate model might well be used in an iterative procedure to design a polyarylate surface that maximizes FRLF NMA. In this process, surrogate models would be used to predict the target property (or properties) for a larger set of hypothetical polymers within the same polymer space from which only the most promising candidates would proceed to chemical synthesis and experimental testing. Results from the experiments would then be "recycled" to validate and, if needed, refine the surrogate model. In fact, this paradigm represents the essence of a dynamic data-driven computational technology for the iterative structural refinement and design of polymeric biomaterials currently underway in our laboratory.

7.7. CONCLUSIONS AND FUTURE PROSPECTS

In the course of our studies, we obtained datasets with hundreds of individual data points. The most intuitive way to analyze such large datasets is by visual inspection of appropriately arranged graphic presentations; however, complex structure–function–performance relationships are easily missed in this approach of data analysis. Alternatively, the evaluation of data for a family of structurally related polyarylates using computational models allowed for the identification of additional correlations between chemical structure and biological response. Of course, trends in the data are expected, but the number of ways a computer can "describe" a molecule on the basis of its intrinsic structure, termed *descriptors*, are beyond those usually considered by most chemists, such as log of the water–octanol partition coefficient and number of hydrogen atoms per repeat unit. These types of correlations potentially can explain the inexplicable. Why does air–water contact angle, a measure of hydrophobicity of a substrate, often strongly correlate with cell proliferation in some cases and exhibit virtually no correlation in other studies? Our studies revealed that hydrophobic substrates significantly decreased cell proliferation except when the carbon atoms were replaced with oxygen atoms, in which case the substrate actually supported cell proliferation. In this case, both materials exhibited similar hydrophobicity, but the differences at the molecular level resulted in distinctly different bioresponses. Remarkably, the intricate relationship of hydrophobicity, molecular structure, and cell proliferation was accurately predicted using a computational model, which was fine-tuned to the parameters, or descriptors, that were essential for cell proliferation [80]. This approach would allow for the rational design of optimized biomaterials in record time, providing specialty materials with optimized chemical, physicomechanical, and biological properties [1,6,82].

Additionally, Reynolds demonstrated a reasonable predictive model of glass transition temperature and air-water contact angle parameter space could be developed based on experimental data for only 17 of 112 polymers, which were representative of polyarylate structural space [37]. This is not only a valuable example of how experimental time can be saved, but an important demonstration of how rational approaches can be used to identify particular candidates amidst an overwhelming number of combinations. Appropriate design of experiment is crucial to accelerated discovery of new structure–function relationships and new biomaterials. First, diverse libraries can be designed to represent the entire virtual library for initial screening of parameter space [37]. In a second iteration, a more focused library can be designed around the initial "hits" of the first library to fine-tune the formulation. The ability to perform a large number of experiments allows for the development of quantitative structure–property relationships (QSPRs), which can be used for the prediction of optimized materials. This iterative cycling of experimental results with computational models results in the best results in the least time [83]. These models will be discussed in detail in the following section. As the database of structure–property–performance is increased for different types of polymer, it is feasible that virtual libraries of polymers will be evaluated based solely on the structure of their molecular repeat unit and their structural similarity to other polymers in the database. This type of technology would truly result in the rational design of biomaterials, and an accelerated rate of discovery, saving hours of research time and dollars.

Our extensive studies have demonstrated success in adapting computational approaches routinely employed in pharmaceutical drug design to offer guidance and direction in the design, selection, and optimization of novel biorelevant materials. This paradigm works quite well for the case studies considered here involving a combinatorial library of polymers and the "target" properties of protein adsorption and cell proliferation. However, the methodology is sufficiently general to treat many other problems in biological response. Since the time and expense in computing polymer descriptors are far less than in vitro or in vivo measurements, computational modeling approaches such as described here can significantly reduce the costs and labor associated with developing high-performance biomaterials for specific applications. It is hoped that efforts will inspire materials scientists to incorporate computational approaches in their discovery programs.

REFERENCES

1. Castner, D. and Ratner, B., Biomedical surface science: Foundations to frontiers, *Surf. Sci.* **500**:28–60 (2002).
2. Hench, L. and Polak, J., Third-generation biomedical materials, *Science* **295**:1014–1017 (2002).
3. Peppas, N. A. and Langer, R., New challenges in biomaterials, *Science* **263**:1715–1720 (1994).
4. Ratner, B. D., Castner, D. G. et al., Biomolecules and surfaces, *Vacuum Sci. Technol. A* **8**(3, Pt. 2):2306–2317 (1990).
5. Hubbell, J., Biomaterials in tissue engineering, *Biotechnology* **13**:565–576 (1995).
6. Kohn, J., New approaches to biomaterials design, *Nature Mater.* **3**:745–747 (2004).
7. Kulkarni, R. K., Pani, K. C. et al., Polylactic acid for surgical implants, *Arch. Surg.* **93**:839–843 (1966).
8. Törmälä, P., Vasenius, J. et al., Ultra-high strength absorbable self-reinforced polyglycolide (SR-PGA) composite rods for internal fixation of bone fractures: *In vitro* and *in vivo* study, *J. Biomed. Mater. Res.* **25**:1–22 (1991).
9. Ray, J. A., Doddi, N. et al., Polydioxanone (PDS), a novel monofilament synthetic absorbable suture, *Surg. Gynecol. Obstet.* **153**:497–507 (1981).
10. Chasin, M., Domb, A. et al., Polyanhydrides as drug delivery systems, in *Biodegradable Polymers as Drug Delivery Systems*, M. Chasin and R. Langer (eds.), New York, NY, Marcel Dekker, New York, 1990, pp. 43–70.
11. Abramson, S., Kohn, J. et al., Bioresorbable and bioerodible materials, in *Biomaterials Science*, B. Ratner, A. Hoffman, F. Schoen, and J. Lemons (eds.), Academic Press, San Diego, 2004.
12. Menger, F. M., Eliseev, A. V. et al., Phosphatase catalysis developed via combinatorial organic chemistry, *J. Org. Chem.* **60**:6666–6667 (1995).
13. Holder, A. J. and Kilway, K. V., Rational design of dental materials using computational chemistry, *Dent. Mater.* **21**:47–55 (2005).

14. Webster, D. C., Bennett, J. et al., High throughput workflow for the development of coatings, *J. Coat. Technol.* **1**(6):34–39 (2004).

15. Meier, M. A. R., Hoogenboom, R. et al., Combinatorial methods, automated synthesis and high-throughput screening in polymer research: The evolution continues, *Macromol. Rapid Commun.* **25**:21–33 (2004).

16. Kemnitzer, J. and Kohn, J., Degradable polymers derived from the amino acid L-tyrosine, in *Handbook of Biodegradable Polymers*, A. J. Domb, J. Kost, and D. M. Wiseman (eds.), Harwood Academic Publishers, Amsterdam, 1997, Vol. 7, pp. 251–272.

17. Brocchini, S., James, K. et al., A combinatorial approach for polymer design, *J. Am. Chem. Soc.* **119**(19):4553–4554 (1997).

18. Brocchini, J., James, K. et al., Structure-property correlations in a combinatorial library of degradable biomaterials, *J. Biomed. Mater. Res.* **42**:66–75 (1998).

19. Smith, J. R., Seyda, A. et al., Integration of combinatorial synthesis, rapid screening, and computational modeling in biomaterials development, *Macromol. Rapid Commun.* **25**:127–140 (2004).

20. Moore, J. S. and Stupp, S. I., Room temperature polyesterification, *Macromolecules* **23**(1):65–70 (1990).

21. Yu, C. and Kohn, J., Tyrosine-PEG-derived poly(ether carbonate)s as new biomaterials. Part I: Synthesis and evaluation, *Biomaterials* **20**(3):253–264 (1999).

22. Tziampazis, E., Kohn, J. et al., PEG-variant biomaterials as selectively adhesive protein templates: Model surfaces for controlled cell adhesion and migration, *Biomaterials* **21**:511–520 (2000).

23. Sharma, R. I., Kohn, J. et al., Poly(ethylene glycol) enhances cell motility on proteinbased poly(ethylene glycol)–polycarbonate substrates: A mechanism for cell-guided ligand remodeling, *J. Biomed. Mater. Res.* **69A**:114–123 (2004).

24. Winblade, N. D., Schmokel, H. et al., Sterically blocking adhesion of cells to biological surfaces with a surface-active copolymer containing poly(ethylene glycol) and phenylboronoic acid, *J. Biomed. Mater. Res.* **59**(4):618–631 (2002).

25. Kohn, J., Bolikal, D. et al., *Radio-opaque Polymer Biomaterials*, Rutgers University, New Brunswick, NJ 1998.

26. Zeltinger, J., Schmid, E. et al., Advances in the development of coronary stents, *Biomater. Forum* (Official Newsletter of the Society for Biomaterials) **26**(1):8–24 (2004).

27. Brandom, D. K., Schmid, E. et al., *Inherently Radiopaque Polymeric Products for Embolotherapy*, PCT Int. Appl., Rutgers University, 82, 2005.

28. Ertel, S. I. and Kohn, J. Evaluation of a series of tyrosine-derived polycarbonates for biomaterial applications, *J. Biomed. Mater. Res.* **28**:919–930 (1994).

29. James, K. and Kohn, J., Applications of pseudo-poly(amino acid) biomaterials, *Trends Polym. Sci.* **4**(12):394–397 (1996).

30. Mori, Y., Nagaoka, S. et al., A new antithrombogenic material with long polyethylene oxide chains, *Trans. Am. Soc. Artif. Intern. Org.* **28**:459–463 (1982).

31. Abramson, S. D., Bolikal, D. et al., Small changes in polymer structure can dramatically increase degradation rates: The effect of free carboxylate groups on the properties of tyrosine-derived polycarbonates, *Trans. 6th World Biomaterials Congress*, Kamuela, HI (USA), Society for Biomaterials, 2000.

32. Tamai, H., Igaki, K. et al., Initial and 6-month results of biodegradable poly(L-lactic acid) corononary stents in humans, *Circulation* **102**(4):399–404 (2000).
33. Baker, W. and Sansbury, H., Preparation of iodine-containing x-ray contrast substances. II. alpha-phenyl-beta-(3,5-diiodo-4-hydroxyphenyl)proprionic acid ("Biliselectan"), *J. Soc. Chem. Ind.* **62**:191–192 (1943).
34. Windholz, M., Budavari, S. et al. (eds.), *The Merck Index*, Merck & Co, Rahway, NJ, 1983.
35. Harington, C. R. and Randall, S. S., Observations on the iodine-containing compounds of the tyroid gland. Isolation of DL-3,5-diiodotyrosine, *Biochem. J.* **23**:373–383 (1929).
36. Kohn, J., Weber, N. et al., Biomaterials informatics: Data sets for biomaterials modeling obtained by a combinatorial approach, paper presented at 19th European Conf. Biomaterials, Sorrento, Italy, 2005.
37. Reynolds, C. H., Designing diverse and focused combinatorial libraries of synthetic polymers, *J. Combin. Chem.* **1**(4):297–306 (1999).
38. Belu, A. M., Brocchini, S. et al., Characterization of combinatorially designed polyarylates by time-of-flight secondary ion mass spectrometry, *Rapid Commun. Mass Spectrom.* **14**:564–571 (2000).
39. Zhang, H., Hoogenboom, R. et al., Combinatorial and high-throughput approaches in polymer science, *Meas. Sci. Techonol.* **16**:203–211 (2005).
40. Meredith, J. C., Sormana, J. L. et al., Combinatorial characterization of cell interactions with polymer surfaces, *J. Biomed. Mater. Res.* **66A**(3):483–490 (2003).
41. Anderson, D. G., Levenberg, S. et al., Nanoliter-scale synthesis of arrayed biomaterials and application to human embryonic stem cells, *Nature Biotechnol.* (published online June 13, 2004).
42. Weber, N., Bolikal, D. et al., Small changes in the polymer structure influence the adsorption behavior of fibrinogen on polymer surfaces: Validation of a new rapid screening technique, *J. Biomed. Mater. Res.* **68(A)**:496–503 (2004).
43. Anderson, D. A., Putnam, et al., D. Biomaterial microarrays: Rapid, microscale screening of polymer-cell interaction, *Biomaterials* **26**:4892–4897 (2005).
44. Suh, K. Y., Khademhosseini, A. et al., Patterning and separating infected bacteria using host-parasite and virus-antibody interactions, *Biomed. Microdevices* **6**(3):223–229 (2004).
45. Ryan, P. L., Foty, R. A. et al., Tissue spreading on implantable substrates is a competitive outcome of cell-cell vs. cell-substratum adhesivity, *Proc. Natl. Acad. Sci. USA* **98**(8):4323–4327 (2001).
46. Magnani, A., Peluso, G. et al., Protein adsorption and cellular/tissue interactions, in *Integrated Biomaterials Science*, Kluwer Academic/Plenum, New York, 2002, pp. 669–689.
47. Bloom, A. L., Forbes, C. D. et al., *Haemostasis and Thrombosis*, Livingstone, Edinburgh, 1994.
48. Mulzer, S. R. and Brash, J. L., Analysis of proteins adsorbed to glass from human plasma using immunoblotting methods, *J. Biomater. Sci. Polym. Ed.* **1**:173–182 (1990).
49. Sigal, G. B., Mrksich, M. et al., Effect of surface wettability on the adsorption of proteins and detergents, *J. Am. Chem. Soc.* **120**:3464–3473 (1998).

50. Hammon, B. D., Pollack, B. A. et al., Binding of a Pleckstrin homology domain protein to phosphoinositide in membranes: A miniaturized FRET-based assay for drug screening, *J. Biomed. Screen.* **7**:45–55 (2002).

51. Weber, N., Wendel, H. P. et al., Formation of viscoelastic protein layers on polymeric surfaces relevant to platelet adhesion, *J. Biomed. Mater. Res.* **72A**:420–427 (2005).

52. Abramson, S. D., Alexe, G. et al., A computational approach to predicting cell growth on biomaterials, *J. Biomed. Mater. Res.* **73A**:116–124 (2005).

53. Abramson, S., Smith, D. et al., Non-gaussian distributions can invalidate common statistical methods in analyzing cell culture experiments, paper presented at Society for Biomaterials 29th Annual Meeting, Reno, NV, Society for Biomaterials, 2003.

54. Dalby, M. J., Riehle, M. O. et al., Polymer-demixed nanotopography: control of fibroblast spreading and proliferation, *Tissue Eng.* **8**:1099–1108 (2002).

55. Washburn, N. R., Yamada, K. M. et al., High-throughput investigation of osteoblast response to polymer crystallinity: Influence of nanometer-scale roughness on proliferation, *Biomaterials* **25**:1215–1224 (2005).

56. Voet, D. and Voet, J. G., *Biochemistry*, Wiley, New York, 1990.

57. Heid, C. A., Stevens, J. et al., Real time quantitative PCR, *Genome Res* **6**(10):986–994 (1996).

58. Lockey, C., Otto, E. et al., Real-time fluorescence detection of a single DNA molecule, *Biotechniques* **24**(5):744–746 (1998).

59. Schmid-Kotsas, A., Gross, H.-J. et al., Lipopolysaccharide-activated macrophages stimulate the synthesis of collagen type I and c-fibronectin in clutured pancreatic stellate cells, *Am. J. Pathol.* **155**(5):1749–1758 (1999).

60. Hoogenboom, R., Meier, M. A. R. et al., Combinatorial methods, automated synthesis and high-throughput screening in polymer research: Past and present, *Macromol. Rapid Commun.* **24**(1):15–32 (2003).

61. Potyrailo, R. A. and Takeuchi, I. E., Special issue on combinatorial and high-throughput materails research, *Meas. Sci. Techonol.* **16**(1) (2005).

62. Adams, N. and Schubert, U. S. Software solutions for combinatorial and high-throughput materials and polymer research, *Macromol. Rapid Commun.* **25**:48–58 (2004).

63. Effah-Kaufmann, E. A. B. and Kohn, J., Correlations of osteoblast activity and chemical structure in the first combinatorial library of degradable polymers, *Trans. 6th World Biomaterials Congress*, Kamuela, HI (USA), Society for Biomaterials, 2000.

64. Spivak, J. M., Blumenthal, N. C. et al., A new canine model to evaluate the biological effects of implant materials and surface coatings on intramedullary bone ingrowth, *Biomaterials* **11**(1):79–82 (1990).

65. Suganuma, J. and Alexander, H., Biological response of intramedullary bone to poly-L-lactic acid, *J. Appl. Biomater.* **4**:13–27 (1993).

66. Choueka, J., Charvet, J. L. et al., Canine bone response to tyrosine-derived polycarbonates and poly(L-lactic acid), *J. Biomed. Mater. Res.* **31**:35–41 (1996).

67. Todeschini, R., Mannhold, R. et al., *Handbook of Molecular Descriptors*, Wiley, New York, 2000.

68. Chemical Computing Group Inc., MOE (The Molecular Operating Environment), Montreal, Canada H3A 2R7, 2003.
69. Todeschini, R., Consonni, V. et al., Dragon Web version, Milano, Italy, 2003.
70. Hertz, J., Palmer, R. G. et al., *Introduction to the Theory of Neural Computation*, Addison-Wesley, Redwood City, CA, 1991.
71. Cristianini, N. and Shawe-Taylor, J., *An Introduction to Support Vector Machines*, Cambridge University Press, Cambridge, UK, 2000.
72. Draper, N. and Smith, H., *Applied Regression Analysis*, 2nd ed., Wiley, New York, 1981.
73. Rasheed, K., Hirsh, H. et al., A genetic algorithm for continuous design space search, *Artif. Intell. Eng.* **11**(3):295–305 (1997).
74. Research, P. L. R., C5.0. St Ives NSW 2075, Australia, 2002.
75. Mitchell, T. M., *Machine Learning*, McGraw-Hill, New York, 1997.
76. Wold, H., Partial least squares, in *Encyclopedia of Statistical Sciences*, S. Kotz and N. Johnson (eds.), Wiley, New York, 1985, 6:581–591.
77. Joliffe, I., *Principal Component Analysis*, Springer-Verlag, New York, 1986.
78. Geladi, P. and Kowalski, B., Partial least squares regression: A tutorial, *Analytica Chimica Acta* **185**:1–17 (1986).
79. Smith, J., Knight, D. et al., Using surrogate modeling in the prediction of fibrinogen adsorption onto polymer surfaces, *J. Chem. Inform. Comput. Sci.* **44**:1088–1097 (2004).
80. Kholodovych, V., Smith, J. et al., Accurate predictions of cellular response using QSPR: A feasibility test of rational design of polymeric biomaterials, *Polymer* **45**:7367–7379 (2004).
81. Nair, A., Jayatilleke, P. et al., Computational studies on tetrahydropyrimidine-2-one HIV-1 protease inhibitors: improving three-dimensional quantitative structure-activity relationship comparative molecular field analysis models by inclusion of calculated inhibitor- and receptor-based properties, *J. Med. Chem.* **45**(4):973–983 (2002).
82. Ratner, B. D., The engineering of biomaterials exhibiting recognition and specificity, *J. Mol. Recognit.* **9**(5–6):617–625 (1996).
83. Schmatloch, S., Bach, H. et al., High-throughput experimentation in organic coating and thin film research: State-of-the-art and future perspecitives, *Macromol. Rapid Commun.* **25**:95–107 (2004).

CHAPTER 8

Combinatorial Methods and Their Application to Mapping Wetting–Dewetting Transition Lines on Gradient Surface Energy Substrates[1]

KAREN M. ASHLEY and D. RAGHAVAN
Polymer Group
Department of Chemistry
Howard University
Washington, DC

AMIT SEGHAL, JACK F. DOUGLAS, and ALAMGIR KARIM
Polymers Division
National Institute of Standards and Technology (NIST)
Gaithersburg, Maryland

8.1. INTRODUCTION

The development and application of combinatorial methods and techniques to pharmaceutical drug discovery has revolutionized the process of bringing new drugs to market [1,2]. Similarly, the application of combinatorial principles to material science is expected to greatly influence the future direction of materials research [3–5]. The combinatorial method has played a key role in the discovery and optimization of novel inorganic and organic materials [6–12]. One area where high-throughput techniques have made a big impact is catalysis. Another area of materials science where these principles are just beginning to be applied is in identifying new compositions and/or mapping of

[1] Official contribution of the National Institute of Standards and Technology; not subject to copyright in the United States.

Combinatorial Materials Science, Edited by Balaji Narasimhan, Surya K. Mallapragada, and Marc D. Porter

processing parameters that influence polymer properties [13–16]. The particular utility of combinatorial methods in studying complex polymer materials including coatings, structural plastics, personal-care products, biomaterials, and nanostructured materials has been recognized. Some of the complexities of these systems arise from reactions, interfacial phenomena, and transport behavior that occur during materials synthesis and processing. These phenomena can be influenced by temperature, polymer type, solvent, annealing conditions (time and pressure), and film thickness.

Combinatorial techniques provide a new paradigm for investigating complex phenomena involving multiple variables by executing many experiments in parallel. Four-basic steps are involved in the process: (1) experimental design, (2) creation of sample "libraries," (3) high-throughput screening, and (4) data analysis or "informatics." Combinatorial techniques have been applied for the optimization of reaction parameters (time, temperature, catalysts) for synthesis of specific polymers, library creation of surface grafted polymers and graded polymer films, mechanical testing of polymers, coating formulation and testing, and understanding film stability [13–29]. Our discussion is limited to the use of the combinatorial methodology to study the physics of dewetting–wetting behavior of thin polymer films on controlled energy surfaces, while the reader is referred to read a review by Takeuchi et al. [30] for obtaining more information about the successful examples where combinatorial techniques have made an impact not only in polymer synthesis and characterization but also in electronic, magnetic, and catalytic materials.

The dewetting behavior of thin film is an ideal subject to be studied by combinatorial and other high throughput methods because the number of parameters that influence the kinetic and thermodynamic behavior is very large in general. For example, film wettability depends on molecular mass, temperature, the interfacial energies describing the polymer–substrate and polymer–air interactions [31–34]. Identifying the thermodynamic conditions under which films are wettable is highly relevant because many of the contemporary coating technologies require uniform and stable thin polymeric films for use in scientific and technological applications such as adhesives, nanodevices, sensors, lithography, lubricants, biomaterials, and paints.

The breadth of parameter space and the complex interaction between these variables poses a challenge for investigation by conventional experimental techniques, which often rely on manipulating a single parameter over several experiments. Combinatorial approaches can be highly productive and cost-efficient. Combinatorial approaches are designed to generate enormous amount of experimental data over a broad range of variables over a relatively short time period.

While the use of combinatorial technique in the synthesis and characterization of novel polymers may seem relatively straightforward, there are several hurdles that need to be cleared before its adaptation to the field of polymer science. Unfortunately, the rapid adaptation of combinatorial techniques once devised for drug discovery to the field of polymer science has met a lot of roadblocks, especially in the design of appropriate libraries of measurements [35,36]. Many of the morphological properties of the continuous polymer libraries are obtained using optical and/or atomic force microscopy. Since the measurement of properties in a combinatorial library change over a small length scale, it is important that the bulk polymer characterization tools such as microscopy be adapted to cover the broad breadth of parameter space and it is also necessary to have the lateral resolution to map localized regions where interesting behavior emerges. Here, we focus on the creation of libraries as it relates to the study of polymer thin film wettability. This choice serves as an illustration of the methodology of a problem of great practical interest in its own right. In particular, we investigate thin polymer films on a substrate having a continuous gradient in surface energy, as well as gradients in temperature, composition, and film thickness. We also investigate a range of polymer molecular masses, where this variable is not treated using gradient methods.

8.2. SURFACE ENERGY GRADIENTS (γ GRADIENT)

A considerable volume of research has been directed to understanding the structure and properties of thin polymeric films on particular substrates for both scientific and technological applications [37,38]. While most experimental studies so far have dealt with wettability of thin liquid films on chemically "homogeneous" substrates, the influence of underlying substrates surface energy gradient on film stability over a broad range has not been systematically considered, despite its importance in developing durable coatings. There have been few exceptions, where comparative studies of polymer film dewetting on two or more discrete substrates with different surface energies have been performed [39,40]. Gradient energy surfaces can be prepared by chemical etch method or UVO treatment of chlorosilane SAM (self-assembled monolayer) treated Si substrates and provide the foundation for the current study.

8.2.1. Chemical Etch Method

The experimental technique used to change the chemical nature of the solid Si surface was buffered oxide etching followed by an oxidizing "Piranha" ($H_2SO_4/H_2O_2/H_2O$) etch [41,42]. The gradient etching procedure involves dipping an "as received" silicon wafer (Polishing Corporation of America[3]) in an aqueous HF/NH_4F buffered oxide etch (J. T. Baker) for 3 min followed by 5 min in a volume fraction 40% NH_4F aqueous solution (J. T. Baker) to

[3] Certain commercial instruments and materials are identified in this chapter to adequately describe the procedure. In no case does such identification imply recommendation or endorsement by the National Institute of Standards and Technology, nor does it imply that the instruments or materials are necessarily the best available for the purpose.

produce a hydrophobic SiH terminated substrate. The hydrophobically modified Si wafer was immersed at controlled rates over a large distance into a Piranha solution (30 vol % H_2SO_4 / 70 vol % H_2O_2) at 80 °C using a motion stage (CompuMotor). Figure 8.1a shows schematically the immersion of silicon wafer in a Piranha solution at a constant rate (0.1 to 2 mm/s) over a distance of 30 mm. The entire substrate was then rapidly withdrawn from the solution, rinsed with deionized (DI) water. The gradual immersion of substrate into the Piranha etch solution creates a gradient in the substrate surface energy. The extent of surface oxidation is sensitive to the composition of the Piranha etch solution, temperature of the Piranha etch solution bath, and the immersion rate of the substrate into the Piranha etch solution. Figure 8.1b is a schematic representation of the chemical reactions that occur on the silicon wafer on exposure to the buffer and Piranha etch solution. The mechanism describing HF and Piranha etch solution reaction of Si wafer is comprehensively described in Ref. 43. Because of a higher chemical conversion of Si/Si—H to SiO_x/SiOH,

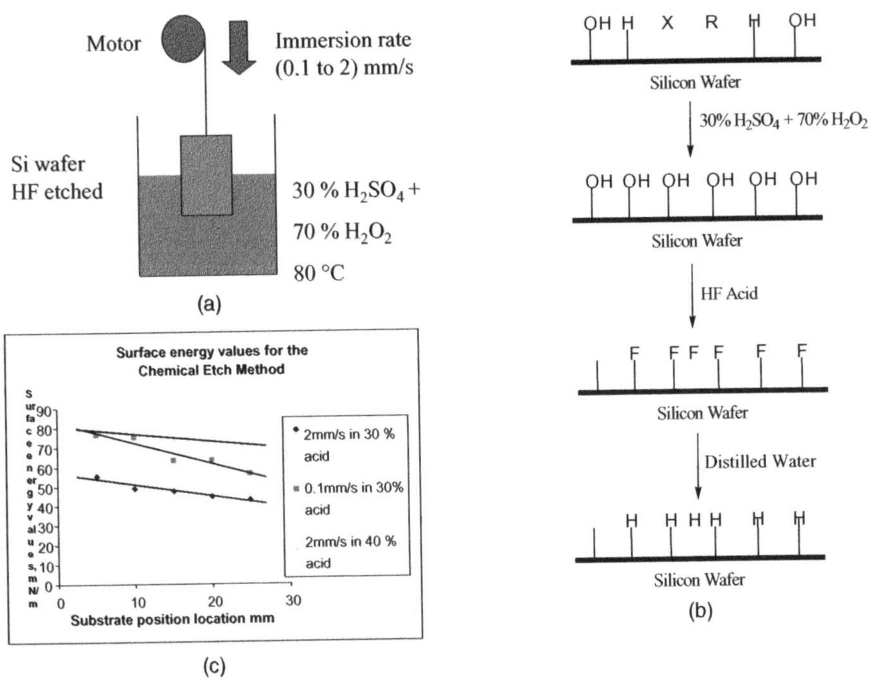

Figure 8.1. (a) Pictorial representation of the immersion of silicon wafer in a Piranha etch solution at a constant rate (0.1 to 2 mm/s) over a distance of 30 mm; (b) chemical reactions that occur on the silicon wafer on exposure to buffer solution and Piranha etch solution; (c) plot of surface energy of the substrate as a function of substrate position.

and the growth of a thicker oxide layer on longer exposure to the Piranha etch solution, a more hydrophilic surface is formed. Figure 8.1c shows the surface energy of the substrate as a function of substrate position. Longer exposures of substrate to Piranha etch solution and higher concentrations of H_2SO_4 in the Piranha etch solutions led to more hydrophilic substrate. This approach provides a method to vary the strength of polar interaction on a single test substrate. AFM indicates a roughened surface indicative of etching agent altering the smooth characteristics of flat silicon substrate. Because of the roughness of the underlying substrate, the task of evaluating shallow gradient in surface chemistry over a large range of surface energy and film thickness on polymer film stability may be rather complicated. Using this approach, 3-dimensional combinatorial studies (e.g., surface energy, polymer film thickness, roughness of underlying substrate) could be made that would have otherwise required orders of magnitude more time.

8.2.2. UVO Treatment of Chlorosilane-Treated Si Substrates

Substrate libraries with gradients in surface energy were obtained by ozonolysis of a chlorosilane SAM-treated silicon surfaces by exposing the SAM layer to ultraviolet light with use of a custom-designed gradient neutral density filter (Maier Photonics with vapor-deposited inconel of Ni, Cr, and Fe alloy, and an optical density of 0.04–1.0 in 11 steps), as described by Robertson et al. [44]. Figure 8.2a shows the assembly of the linear density spacer mounted on glass holders over the wafer and placed in the UV/ozone chamber. The linear density filter controls the amount of illumination of UV light incident on the substrate. A systematic decrease in the water contact angle of the substrate was noticed as a function of substrate position, depending on the UV exposure. Figure 8.2b is a plot of substrate surface energy and water contact angle. The variation of the surface energy of the sample reflects the variation in polar component of the surface energy measurement, while the dispersive component remained nearly constant. Apparently, the dispersive component corresponding to the van der Waals interaction is nearly constant because the underlying silicon substrate has a constant surface energy, and SAM chains assembled are generally of equal length, while the polar component varies with the extent of exposure of the substrate to UVO treatment due to terminal group conversion [44]. UVO treatment causes not only bond breaking but also crosslinking of the silanized monolayers. The functionality of the terminal methyl groups of the monolayer can be changed to primarily carboxylic acids depending on the dosage and duration of exposure. The duration of exposure lasted for 3–5 min, depending on the surface energy range desired.

UVO treatment of monolayer provides an easy method to create gradient surface energy on a single substrate, and provides a natural approach to decouple the effect of polar surface energy component of the substrate on PS film dewetting. Importantly, this method eliminates complications from substrate roughness to overall film stability, as verified by AFM. Surface energy

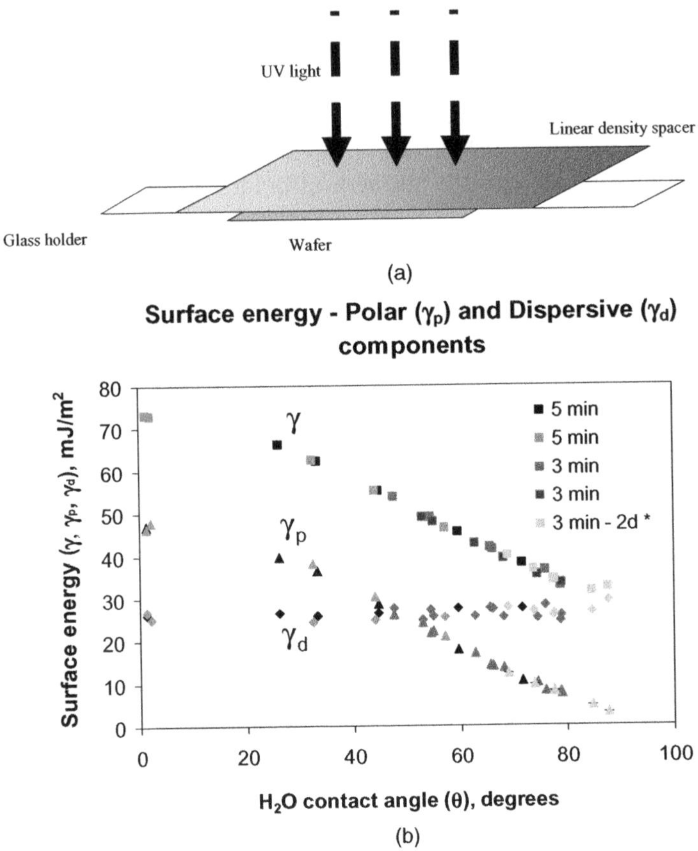

Figure 8.2. (a) Assembly of the linear density spacer mounted on glass holders over the wafer and placed in the UV ozone chamber; (b) Plot of surface energy as a function of water contact angle of the substrate.

gradients prepared by this method were relatively more stable than that prepared by chemical etch method.

8.2.3. Compositional Gradient of SAM on Au Substrates

Another procedure for varying substrate energy on a single substrate is based on composition gradient of SAM [45,46]. In this procedure, alkane thiolates with COOH and CH_3 terminal groups are allowed to diffuse from opposite ends of polysaccharide matrix deposited on the top of a gold substrate. The rate of alkane thiolate diffusion creates a concentration gradient between the two types of thiolates that ultimately results in a substrate with a surface energy gradient. Similar to the UVO treatment of SAM layer, this method allows one to endow solid surfaces with a systematic variation in polar surface energy component while keeping the surface flat on a molecular scale. This method differs from UVO treatment of SAM layer in generating a continuous surface energy gradient instead of a step-gradient in the substrate surface energy.

8.2.4. Characterization of Gradient Energy Substrates

These approaches of preparing shallow gradient in surface chemistry over a wide range of surface energy across large (square centimeters) substrate demand the use of appropriate chemical characterization technique to measure surface energy at a spot with reasonable accuracy. To estimate the surface energy of the substrate, spatially resolved static contact angles of water and diiodomethane are obtained with the use of contact angle goniometer. A minimum of four measurements is recorded for each liquid at any point along the various regions of the gradient energy substrate. The total surface energy of the substrate (sum of the polar and dispersive component) was estimated using the geometric means approach of Owens and Wendt [47]. In this approach, calculations of the polar and nonpolar or dispersion force components of the material are based on contact angle data and the polar and nonpolar components of surface tension for the two liquids. Surface tension values of 72.8 mJ/m^2 and 50.8 mJ/m^2 for water and methylene iodide, respectively, and their respective polar component values of 50.7 mJ/m^2 and 1.8 mJ/m^2 were used for the calculation of substrate surface energy (48).

8.3. TEMPERATURE GRADIENT LIBRARIES (*T* GRADIENT)

The high-throughput nature of measurements on the gradient energy substrates can be augmented by applying thermal gradient orthogonal to the substrate gradient properties to obtain maps of the different substrate state and annealing temperature. Figure 8.3a is a schematic of combinatorial cross-gradients of surface energy and temperature. The thermal gradient is created by placing the polymer coated substrate on a preheated aluminum stage. One edge of the aluminum stage has a higher temperature from a metal heating rod whose heating is controlled by a thermal controller and the opposite edge has a lower temperature whose temperature is controlled by circulation of fluid from a refrigerating bath. Figure 8.3b is a pictorial representation of the wafer and film placed on the temperature gradient aluminum stage. With careful manipulation of the heat and cooling settings, the appropriate temperature limits for the measurements and the linear gradient in temperature across the stage are achieved. Constant purging of nitrogen gas stream over the specimen minimizes film oxidation and environmental contamination. Using the appropriate setting, end point temperatures typically ranging from 80 °C to 150 °C could be easily achieved across ≈ 3 cm breadth specimen. Surface temperature measurements are periodically recorded by using a thermocouple to verify the linearity of the *T*-gradient across the substrate.

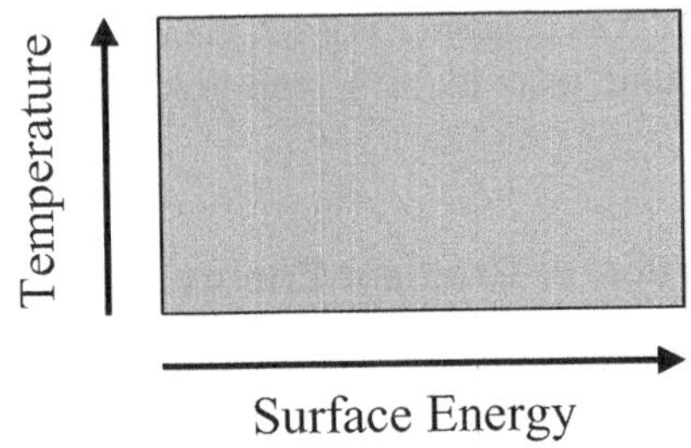

(a)

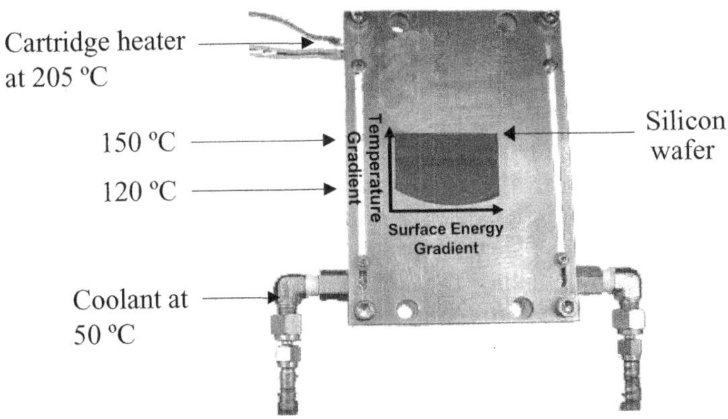

(b)

Figure 8.3. (a) Schematic representation of combinatorial cross-gradients of surface energy and temperature (arrows indicate gradient in different coordinate directions); (b) Assembly of the wafer placed on the temperature gradient aluminum stage.

8.4. THICKNESS GRADIENT LIBRARIES (*h* GRADIENT)

For many coating applications, there is considerable interest in studying the stability of film as a function of film thickness. Thickness gradients of polymer films can be prepared on the gradient surface energy substrate with a velocity gradient knife edge coating apparatus [14,15]. A drop of the polymer solution is spread over the substrate under an angled blade at constant acceleration to yield a thin film with a gradient in thickness orthogonal to the direction of the surface energy gradient. With a careful manipulation of substrate velocity, acceleration, the angular setting of the blade, and the concentration of polymer solution, the thickness of film can be varied from several nm to few μm. The film thickness is measured at different positions on the substrate with a UV-visible interferometer employing a 0.5 mm light spot. The film thickness measurements are repeated and confirmed with a minimum of four combinatorial libraries. The polymer film thickness of gradient specimen ranged from 25 nm to 50 nm across ≈ 3 cm breadth specimen.

Alternatively, a combinatorial study may require the polymer film thickness to be kept constant, while varying the underlying substrate surface energy and annealing temperature. Films of constant thickness are prepared by spin coating polymer solution onto the gradient energy substrate. Films of uniform thickness are prepared on gradient and nongradient chemically modified Si substrate. This generally allows observations of the sample on chemically modified gradient surface to be compared to that of the sample on reference nongradient Si substrates at discreet values of surface energy, to cover the endpoint values of the gradient surface and some intermediate ones as well.

8.5. COMPOSITION GRADIENT LIBRARIES (ϕ GRADIENT)

There are basically three steps in preparing composition gradient films. These steps are gradient mixing, gradient deposition, and film spreading. Details of preparing gradient composition libraries can be found in Ref. 14. Generally, Fourier transform Infrared microspectroscopy in reflectance mode (FTIR-RM) mapping, a noncontact, nondestructive method is used to characterize both qualitatively and quantitatively the chemical composition of the surface [14,26]. The spectral point-to-point mapping of the surface of the film is performed in a grid pattern with computer-controlled stage. The spectra extracted from the map are used for quantitative analysis of the composition.

From the γ, T, ϕ, h libraries, several high-throughput screening methods have been developed for understanding the phase behavior of polymer blends, block copolymers, cell–polymer interactions, and dewetting. Meredith et al. (49) have provided successful examples of this type of gradient methodology in a recent review where the γ, T, ϕ, and h libraries have been used in understanding the phase behavior of polymer blends, cell–polymer interactions and dewetting. Here, the discussion is related to the use of γ, T, and h libraries in formulating an understanding of polymer film dewetting on gradient energy surfaces.

8.6. APPLICATION TO CHARACTERIZING THE STABILITY OF THIN POLYMER FILMS

There are several examples in the literature of the use of combinatorial methods for inorganic and organic materials synthesis, development of structure-processing–property relationships, and formulation of physical models [8–11]. For example, combinatorial methods have been successfully applied to measure biological, chemical and physical properties of polymers over large range of multiparameter space. In 2000, Meredith et al. [14,15]

screened the thin film libraries for dewetting behavior over large range of multi-parameter space. Combinatorial libraries spanning a large *T*, *h*, and time (*t*) range, not only reproduced known dewetting structures and phenomena but also enabled to quantify behavior in different regimes i.e. in heterogeneous nucleated hole dewetting phenomena. Most dewetting studies of thin-polymer films have investigated the processes by which films dewet or wet the substrate by a variety of mechanisms appropriate to the conditions at hand, but there has been little effort to map out the thermodynamic factors that control film stability. Furthermore, there have been no systematic studies of film wettability as a function of substrate chemistry over a significantly wide range of surface energy and temperatrue.

8.7. COMBINATORIAL MAPPING

8.7.1. Combinatorial Mapping of γ-*h* Library of the PS Film

It is clear from many incidental observations that film stability depends on many factors such as molecular mass, temperature, thickness, and the interfacial energies describing the polymer–substrate and polymer–air interactions. Given the multidimensional nature of the parameter space involved, we approached this problem using a combinatorial method in which polymer films of fixed molecular mass and varying thickness were cast on substrates exhibiting shallow gradients in surface chemistry over a wide range of surface energy. Unlike instruments used for characterization of conventional experimentation, combinatorial method often relies on nondestructive and time-efficient techniques for sample library characterization. The libraries are screened for dewetting behavior using a Nikon Optiphot-2 automated optical microscope with a Kodak Megaplus ES 1.0 CCD camera mounted on a trinocular head. The images are stored as 8-bit, (1024 × 1024)-pixel digitized gray-scale data.[4]

Figure 8.4 presents composite optical microscopic images ("cells") of a γ-*h* library of the PS film (Goodyear, $M_w = 1800$ g/mol, polydispersity = 1.19, glass transition temperature = 54 ± 1 °C, where M_w is the (mass average) relative molecular mass) at specified positions on the gradient energy Si substrate prepared by chemical etch method after annealing for 2 h. The γ and *h* for each measurement site is taken as an average γ and *h* over the length scale of measurement site (e.g., microscopic image). We observe dark and bright regions that are indicative of wetted and dewetted regimes respectively in the

[4] We define *ultrathin films* below by the thickness *h* range in which the film properties deviate from their bulk counterparts, but where *h* is larger than the *molecularly thin* range in which van der Waals interactions and the presence of a boundary strongly perturb the fluid structure. In polymer fluids, this thickness regime typically ranges from scales on the order of 100–200 nm (regardless of polymer molecular mass) to a few nm.

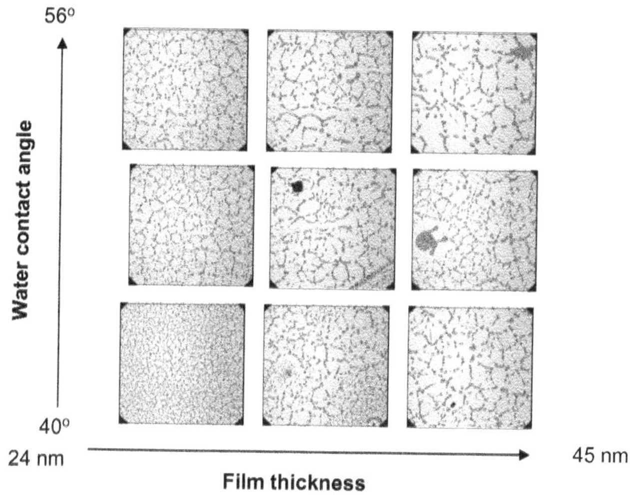

Figure 8.4. Composite optical microscopic images ("cells") of a γ-h library of the PS film (Goodyear, $M_w = 9000\,g/mol$, chain polydispersity = 1.19, glass transition temperature = 54 ± 1 °C, where M_w is the mass average relative molecular mass) at specified position on the gradient energy Si substrate prepared by chemical etch method after annealing for 2h. (Adapted from Ashley et al. [42].)

PS film. Dewetting occurs by the growth of small circular holes that impinge to form small polygons within 2h of annealing time. Close examination of combinatorial γ-*h* libraries at thickness ranging from 24nm to 50nm show distinct dewet patterns that are indicative of differences in film stability across the wafer. Trends in dewetted structure were compared to results reported in literature previously for conventional studies so as to validate the combinatorial method against conventional methodology. The combinatorial libraries reproduce dewetted film structures observed in previous conventional non-combinatorial studies.

We performed a quantitative analysis of the dewetted structures by dividing the combinatorial library into a virtual array of individual cells with various thickness and water contact angles values as the *X* and *Y* coordinates, respectively. By using an automated batch program (NIH Image software), the images were threshold and the number of polygons calculated. By observing regions in the specimen with constant water contact angle where the film thickness increased, it was noticed that the number of polygons in the dewetted polymer film decreased. Our data are in general agreement with literature finding that suggests that when N_p varies as h^{-4}, capillary instability mechanism best describe film dewetting [32,50]. This furthermore validated the usefulness of combinatorial method to study film stability against the "one sample for one measurement" approach.

By observing regions in the specimen with constant film thickness and varying water contact angle, it was noticed that the number density of polygons is larger for the more hydrophilic surface, due to a larger activation of nucleation sites under more unstable conditions [41,42]. At present, these film destabilization effects are not well understood, but it seems clear that a combination of equilibrium (i.e., polymer–surface interactions) and nonequilibrium (destabilizing heterogeneity on surface) conditions are generally involved.

Close examination of combinatorial γ-*h* libraries at water contact angle ranging from 40° to 60° showed that $N_p \sim \theta_{water}^{-3.6 \pm 0.1}$, where θ_{water} is the water contact angle of the substrate and N_p is the number of polygons of dewetted polystyrene film per unit area at constant film thickness. It is pointed out that the films are likely in the crossover from capillary to heterogeneous nucleated regime [15] for the range of film thickness studied [24–50 nm]. Notably, the exponent mentioned above is constant over the polymer film thickness range investigated. This analysis provides the first systematic observation of PS dewetting on a substrates exhibiting shallow gradients in surface chemistry over a range of surface energy and polymer film thickness.

8.7.2. Combinatorial Mapping of γ-T Library of the PS Film

We constructed a combinatorial library of observations for polystyrene film cast on chlorosilane SAM treated Si substrates having an orthogonal surface energy gradient and a gradient in temperature [51,52]. Figure 8.5 presents a collection of appended optical microscopic images of the PS film ($M = 9000$ g/mol) specimen taken at specified position to cover sub-regions (cells) on the gradient surface after annealing the PS film from 94 °C to 152 °C for 50 min. The morphological "film stability map" for 30 nm polystyrene (PS) film showed a curved film stability line governing the transition between wetting and dewetting in ultrathin polymer films as a function of temperature and substrate surface energy. The dewetting–wetting line (DWL) can be seen with the unaided eye as a diffuse boundary separating the stable (wetting) region from the unstable (dewetting) film region. In particular, the DWL (dotted line) shows the film to be stable above and unstable below this line in the temperature–surface energy plane. Interestingly, the wetting–dewetting line is seen to curve in the morphological phase map over a wide lower *T* range, but this curve plateaus at a constant surface energy at high *T*. The diffuse boundary of the DWL line reflects the natural evolution of the microstructure with annealing time and temperature. Although the combinatorial map displayed in this figure was collected after 50 min annealing, similar dewetting trends were noticed even after about 500 min. After an extended annealing time, we notice a minimal shift on the observed DWL line, strongly suggesting that the observed phenomenon is controlled primarily by thermodynamic factors.

An independent confirmation of the transition between the wetting and dewetting regions can be obtained by examining the variation of the droplet

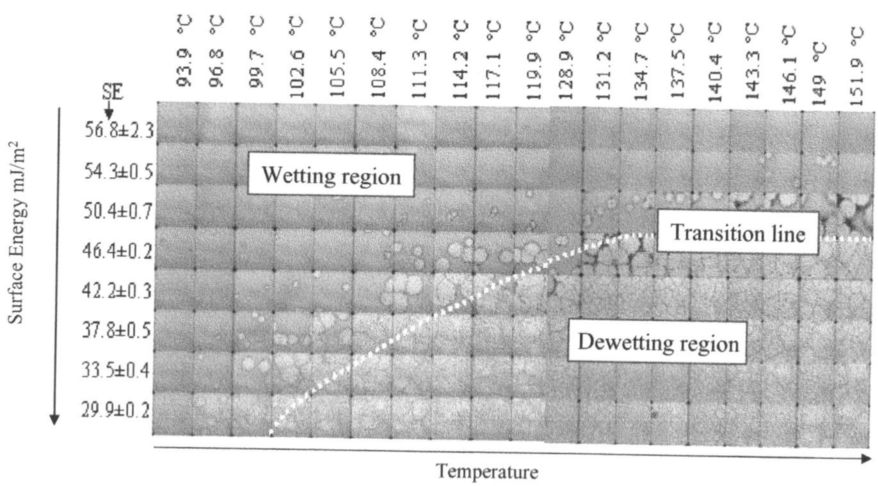

Figure 8.5. Wetting–dewetting transition line obtained by assembling a large number of images of a surface energy–temperature gradient sample. The SE ranges from 28 nN/m to 58 nN/m from top to bottom of the image, while T ranges from 90 °C to 152 °C from left to right end of the image. PS had a relative molecular mass of $M = 9000$ g/mol.

contact angles on polystyrene droplets in the dewetted region as a function of T and E_s. Tapping mode AFM was used to characterize the morphology and contact angle of dewetted polymer film. Figure 8.6 shows representative AFM images of polystyrene dewetting tendency (a) above and (b) below the DWL line. It can be seen that for a fixed dewetting temperature, the PS film is partially wettable (as seen by the droplet configurations) on the lower-surface-energy substrate and wettable on the higher-surface-energy substrate, with occasional limited dewetting holes as shown in Figure 8.6a that do not grow, occurring from heterogeneous nucleation close to the DWL line. This trend in the PS film coverage suggests the existence of a well-defined "dewetting–wetting transition line," as described above. Moreover, one can directly see a transition from wetting to dewetting in the variation of the droplet contact angle as shown in Ref. 51. To observe this progression in the contact angle wetting on approaching the dewetting–wetting transition line (DWL), a planefit filter was applied to all the AFM images. Macros (program) in combination with IGOR analysis was used to automatically identify droplet(s), and take vertical sections of the droplet along the major and minor axes. A minimum of 80 droplets was processed to generate slopes of the section profiles and the height so that the PS contact angle data can be generated. Qualitative analysis of contact angle of dewet droplets below the instability line is consistent with the morphological phase map data. The magnitude of the contact angle of the resulting droplet increases as the substrate energy becomes more unfavorable to wetting. In particular, the system changes from

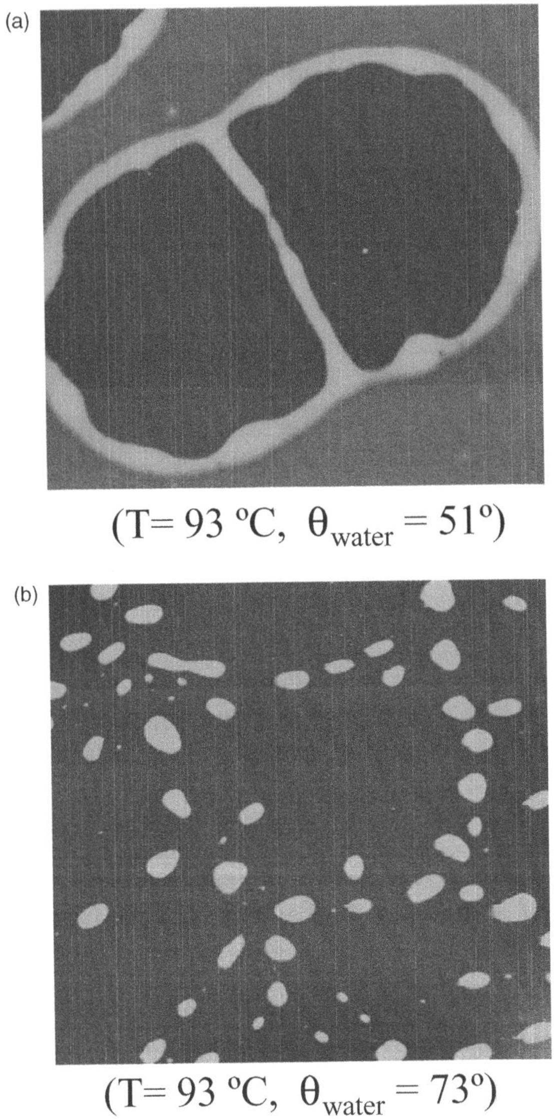

Figure 8.6. AFM images of polystyrene annealed (a) above and (b) below the DWL line. The PS in this image had a relative molecular mass of $M = 1800$ g/mol. Film annealing conditions are listed as (Temperature, Substrate Water Contact Angle).

wetting to dewetting that is accompanied by a change from the zero contact angle situation of the smooth surface to droplets having a finite contact angle. This trend in the contact angle of dewet droplets confirms the existence of a dewetting–wetting transition regime as a function of surface energy of the substrate and temperature.

8.7.3. Combinatorial Mapping of γ-*T*-*M* Library for PS Film

We tried to verify the existence of DWL in ultrathin films for different molecular masses of polystyrene by examining the transition between wetting and dewetting in ultrathin films in the temperature-surface energy plane. Figure 8.7 presents a collection of appended γ-*T* combinatorial maps of the PS film (*M* = 1800, 9000, 15,200, 22,000, and 35,000 g/mol). The different molecular mass values of the polystyrene samples (constant $h \approx 30\,nm$) ranged from oligomeric to entangled molecular mass of polystyrene in the bulk. Repeated examination of a minimum of two to three samples was performed for each molecular mass to verify the reproducibility of the morphological phase map. By following the film evolution over time (up to 9 h), it was found that kinetic effects do not lead to an appreciable variation of the dewetting–wetting transition line location at long times. All the morphological maps showed a curved film stability line governing the transition between wetting and dewetting in ultrathin polymer films as a function of temperature and substrate surface energy as in Figure 8.7. The higher the molecular mass, the more strongly suppressed is the behavior of limited heterogeneous hole formation (due to increased film viscosity) in the wetting regime above the DWL line discussed previously, and hence more sharply defined is the transition.

Although we do not fully understand the origin of this interesting and general feature in the DWL, we can make some observations about this phenomenon and formulate some hypotheses about the origin of this effect that can be checked by measurements. The trend in wetting and dewetting transition line (at, above, and below relative molecular mass M = 9000 g/mol) follows a nearly universal pattern, with the exception of a shift in the knee temperature to higher or lower temperature and the plateau at a higher or lower surface energy depending of polystyrene molecular mass. This led us to consider

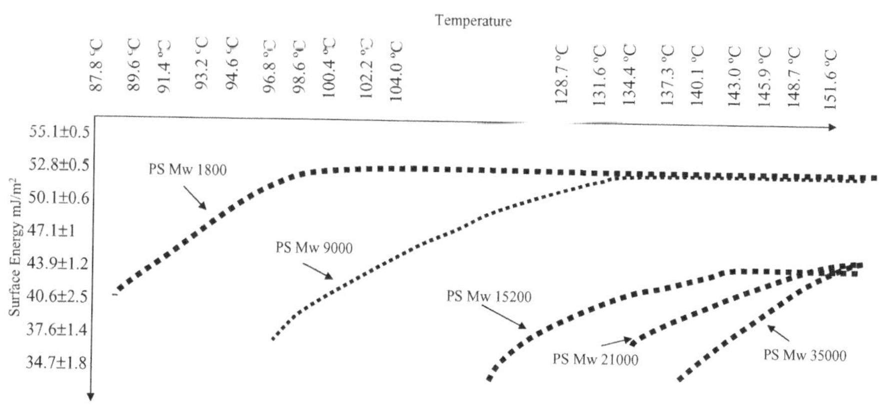

Figure 8.7. Wetting–dewetting transition line obtained for different molecular mass values of polystyrene by assembly from a large number of images of a surface energy–temperature gradient. The PS in this image had relative molecular masses: 1800, 9000, 15,200, 22,000, and 35,000 g/mol.

whether the knee temperature T_{knee} has any correlation with the glass transition temperature T_g of polystyrene. Remarkably, the ratio T_{knee}/T_g was found to be close to 1.5, with the exception of the sample with molecular mass 22,000 g/mol, where $T_{knee}/T_g = 1.6$. A possible link to the glass transition does indeed seem to be suggested by these measurements.

Recent observations [53] on glycerol, another glass-forming liquid, provide insight into why thin film dewetting should exhibit the sensitivity toward glass formation that initiates well above T_g. These measurements indicate that the capillary wave spectrum of glycerol "freezes" at a temperature that roughly equals 1.5 T_g. Since dewetting ultimately has its origin in the unstable growth of thermally excited capillary waves, their suppression by incipient glass formation in thin films would definitely inhibit the dewetting process in these films. These observations provide an attractive possibility for explaining the physical significance of the knee temperature T_{knee} near 1.6 T_g in our PS films. Further studies are needed to understand how glass formation influences polymer film stability and to characterize how the capillary wave spectrum of PS changes in thin films and in the bulk as temperature is lowered toward T_g.

At any rate, our data seem to exhibit a well-defined characteristic temperature and surface energy, and we have thus introduced reduced variables (i.e., reduced surface energy E_s^* and reduced temperature, T^*) to see if we can reduce all of our data to a single master curve. Specifically, we simply normalize E_s by its plateau wetting-dewetting crossover value at high T ($E_s = 46.4$ mN/m in Figure 8.5) and T is reduced by the characteristic temperature at which the wetting–dewetting curve first deviates from a constant E_s ($T_{knee} = 131.2$°C in Figure 8.5). It must be mentioned that for other polystyrene samples, T_{knee} and E_s were found to depend on M. The transition from wetting to dewetting (dewetting–wetting transition line) for PS indeed appears to follow a nearly universal pattern in these reduced variables and this reduction is indicated explicitly in Figure 2 of Ashley et al. [51]. This collapse of our data onto a nearly universal curve was unexpected and it can be inferred that entanglement effects do not significantly modify the film stability because of the potential difficulty of films with higher M to equilibrate. Such a change may occur in much higher M than considered in the present measurements, however.

We now provide a tentative explanation of this remarkable data reduction. Under equilibrium conditions, the film wettability is defined by the spreading coefficient (S). The spreading coefficient ($S > 0$ for nonwettable film) is stated from continuum thermodynamics as

$$S = \gamma_s - \gamma_{sp} - \gamma_p \tag{1}$$

where γ_s and γ_p are the surface free energies of the solid substrate and polymer, respectively, and γ_{sp} is the interfacial free energy between the polymer and the substrate [54,55]. The long-range nature of the substrate interactions is not included in our description of polymer film stability since it is not currently clear how to fundamentally describe the influence of these interactions from the complex substrate on polymer film stability and structure. We note that the substrate surface energies reported are measured values at room temperature and not at the elevated dewetting temperatures, although their temperature dependence is expected to be weak compared to the range of surface energy covered in the typical gradient surface energy substrate. Any reasonable description of the process, however, must capture both these interactions and film confinement. Of course, this approach must break down in the extreme limit of *molecularly thin films*, where the van der Waals interactions have an appreciable influence on fluid structure [55–59]. For molecularly thin films (see footnote 4 above, in Section 8.7.1), we must shift to a description of dewetting in terms of the specific surface potentials involved, based on Lifshitz theory and its various modifications [55,56,58]. Our ultrathin polymer films correspond to an h larger than molecular dimensions, but smaller than the critical thickness required [≈ 100 nm to 200 nm] for the apparent T_g to become insensitive to film thickness [55]. For such films, we can expect the substrate interactions and confinement effects responsible for modifying glass formation to alter thermodynamic properties such as surface tension in a parallel fashion, and we consider this possibility below.

There are technical difficulties in estimating the spreading parameter and γ_s, γ_p, and γ_{sp} as a function of T at elevated temperatures due to the volatility of the test liquids, the possibility of absorption of the test liquid into the thin PS film, and the possible thermal degradation effects of the SAM substrate. We have purposely chosen OTS as our SAM substrate layer since it remains stable to relatively high temperatures, $T \leq 250$ °C [60]. Since γ_s does not change appreciably with T, it is reasonable to speculate that the effect derives either from a change in the T dependence of γ_p and/or γ_{sp}. At present, little is known about the impact of T on the interfacial free energy between the ultrathin polymer film and the underlying substrate (γ_{sp}) and surface energy of polymer (γ_p) for thin films.

The surface tension of PS (γ_p) was not measured at elevated temperature. Instead, we consider as a reference point the *bulk* film γ_p measurements by Dee (M = 1800 g/mol) [61], determined necessarily at relatively elevated T because of the general difficulty of measuring the surface tension of glassy liquids. γ_p changes by an appreciable amount (7 mJ/m^2) for the T range of 50 °C to 150 °C for this relatively low-mass polymer. Notably, previous work has assumed that Dee's bulk data also apply to ultrathin PS films such as our own.

To check whether the surface tension of our ultrathin films is similar to that of macrodroplet surface tension, the surface energy of a solid PS film of varying thickness was estimated at room temperature. It was found that the apparent polymer surface tension depends strongly on film thickness for ultrathin films ($h \leq 100$ nm to 200 nm) [55]. It must be mentioned that the γ_p measurements were made for ultrathin PS cast on a silicon wafer rather than PS macrodroplets obtained from the dewetting of a thick polymer film. Similar changes of the surface tension in ultrathin supported PS films have been reported for a similar thickness range [62]. In these measurements, however, a decrease rather an increase of the apparent surface tension was suggested with increase in film thickness. Since a free film is not subjected to the potential field of the substrate, this is a rather different physical situation to the supported film and thus is not directly comparable. Nonetheless, the commonality of the reported scale at which the finite-size effects show their influence on the apparent surface tension is interesting. These observations together suggest that even the *sign* of the surface tension shift might depend on the physical nature (surface interaction, roughness, supported, or freestanding) of the polymer interface, just as previous reports of the shift of the sign of the apparent T_g in ultrathin films has been claimed to depend on the polymer–surface interaction [63]. It would be interesting to determine whether these phenomena are related.

We argue that there is a weak dependence of γ_p on temperature in ultrathin polymer film; hence the trend observed is a result of a change in the T dependence of γ_{sp}. The interfacial energy of PS (γ_{sp}) was not measured at elevated temperature because of the volatility of the test solvents. Instead, we consider γ_{sp} measurements at ambient temperature by Fryer et al. [64] of ultrathin PS films on an underlying SAM substrate. The interfacial energy γ_{sp} of PS on OTS surface increases with increasing substrate surface energy at ambient temperature.

It is postulated that equilibration of the film must become increasingly difficult as T is lowered toward the glass transition, ultimately leading to an insensitivity of γ_{sp} to T, upsetting the balance between γ_{sp} and γ_s. Specifically, a slower variation of γ_{sp} as T approaches T_g would make a smaller negative contribution to the spreading coefficient (S) so that a *smaller* value of the surface energy would signal the wetting–dewetting transition ($S = 0$), in qualitative consistency with our findings. Such a scenario would imply that the stability of ultrathin films is intimately linked to the physics of glass formation. The tentative interpretation offered to describe the trend is that glass formation can significantly influence the balance of surface energies responsible for the stability of thin films.

As a check on the generality of our arguments above and the form of our data reduction for the dewetting–wetting transition line, we briefly consider data for a different polymeric system. We also considered the hydrophilic biodegradable polymer (PDLA) having a 40nm thickness and an annealing time of 6h on a gradient energy substrate. Libraries of observations for these films were obtained as in the case of the PS films described above. Strikingly, the dewetting–wetting lines found for these PDLA films are remarkably similar to those for the PS films. Given the rather different physical nature of the PS and PDLA polymers (one is an amorphous polymer; the other is liable to crystallization), this suggests that this form of the dewetting-dewetting transition line is general. Obviously further studies are needed to check this universality, but this preliminary observation is very encouraging. The combinatorial method has facilitated this potentially important discovery, which otherwise might have missed in conventional non-combinatorial measurements without the guidance of a predictive theory.

8.8. CONCLUSIONS

Combinatorial methods for preparation of the γ, T, ϕ, and h libraries were described. For creating the γ library across macroscopically large (centimeters) substrates, we present two novel methods (chemical etching of silicon substrate and UV/ozone treatment of chlorosilane SAM-treated silicon substrate) of creating large-surface-energy gradients. Some of these libraries have proved to be extremely useful for understanding the phase behavior of polymer blends, block copolymers, and dewetting. The combinatorial method has not only allowed us to explore and validate known dewetting process, but also helped gain insight into novel regimes that are difficult and time-consuming to explore. Here, we have used combinatorial method in which polymeric films of fixed thickness and molecular mass were cast on substrates exhibiting wide ranges of surface energy and temperature gradient to explore the stability of ultrathin polymer films against dewetting. We observe a near-universal scaling curve describing a dewetting–wetting *transition line* (DWL) as a function of substrate surface energy and temperature for both PS and PDLA films. Tentative explanations are suggested to explain this dewetting–wetting transition phenomenon, and the physics of glass formation is suggested to be relevant to our reduced temperature description of the DWL. Apparently, the glass-formation ability of fluid can significantly influence the balance of surface energies responsible for the stability of thin hydrophilic and hydrophobic polymer films, and this possibility requires further study.

ACKNOWLEDGMENTS

This work was supported by NIST 70NANB1H0060 and NSF DMR-0213695. We are grateful to Eric J. Amis and J. Carson Meredith (Georgia Tech) for contributing ideas on which the present work is based. We are indebted to Michael J. Fasolka and other members of the NIST Combinatorial Methods Center (NCMC, www.nist.gov/combi) for the use of the center facilities for these studies and their indepth help with this project.

REFERENCES

1. Dagani, R., ACS Award in the Chemistry of Materials, *Chem. Eng. News* **76**:66 (2000).

2. Jandeleit, B., Schaefer, D. J., Powers, T. S., Turner, H. W., and Weinberg, W. H., Combinatorial materials science and catalysis, *Angew. Chem. Int. Ed.* **38**:2494 (1999).
3. Xiang, X. D. and Takeuchi, I., *Combinatorial Material Synthesis,* Marcel Dekker, New York, 2003.
4. Amis, E. J., Xiang, X. D., and Zhao, J. C., Combinatorial material science, *MRS Bull.* **27**(4):295 (2002); Amis, E. J., Combinatorial material science, reaching beyond discovery, news & views, *Nature Mater.* **3**:83 (2004).
5. Schubert, U. S. and Amis, E. J. (eds.), Combinatorial research and high-throughput experimentation in polymer and materials research, *Macromol. Rapid Commun.* (special issue) **24**(1) (2003).
6. Wang, J., Yoo, Y., Gao, C., Takeuchi, I., Sun, X., Chang, H., Xiang, X. D., and Schultz, P. G., Identification of a blue photoluminescent composite material from a combinatorial library, *Science* **279**:1712 (1998).
7. Bein, T., Efficient assays for combinatorial methods for the discovery of catalysts, *Angew. Chem. Int. Ed.* **38**:323 (1999).
8. Xiang, X. D., Sun, X., Briceno, G., Lou, Y., Wang, K. A., Chang, H., Wallace-Freedman, W. G., Chen, S. W., and Schultz, P. G., A combinatorial approach to materials discovery, *Science* **268**:1738 (1995).
9. Reddington, E., Spaienza, A., Gurau, B., Vishwanathan, R., Sarangapani, S., Smotkin, E., and Mallouk, T., Combinatorial electrochemistry: A highly parallel, optical screening method for the discovery of better electrocatalysts, *Science* **280**:1735 (1998).
10. Dickinson, T. A., Walt, D. R., White, J., and Kauer, J. S., Generating sensor diversity through combinatorial polymer synthesis, *Anal. Chem.* **69**:3413 (1997).
11. Fodor, S. P. A., Read, J. L., Pirrung, M. C., Stryer, L., Lu, A. T., and Solas, D., Light-directed, spatially addressable parallel chemical synthesis, *Science* **251**:767 (1991).
12. Koinuma, H., Combinatorial materials research projects in Japan, *Appl. Surf. Sci.* **189**:179 (2002).
13. Schmitz, C., Posch, P., Thelakkat, M., and Schmidt, H. W., Efficient screening of materials and fast optimization of vapor deposited OLED characteristics, *Macromol. Symp.* **154**:209 (2000).
14. Meredith, J. C., Karim, A., and Amis, E. J., High throughput measurement of polymer blend phase behavior, *Macromolecules* **33**:5760 (2000).
15. Meredith, J. C., Smith, A. P., Karim, A., and Amis, E. J., Combinatorial material science for polymer thin film dewetting, *Macromolecules* **33**:9747 (2000).
16. Gross, M., Muller, D. C., Nothofer, H. G., Sherf, U., Neher, D., Brauchle, C., and Meerholz, K., Improving the performance of doped π-conjugated polymers for use in organic light-emitting diodes, *Nature* **405**:661 (2000).
17. Wicks, D. A. and Bach, H., The coming revolution for coatings science: High throughput screening for formulations, *Proc. 29th Int. Waterborne High-Solids and Powder Coatings Symp.* 2002.
18. Iden, R., Schrof, W., Hadeler, J., and Lehmann, S., Combinatorial materials research in the polymer industry: Speed versus flexibility, *Macromol. Rapid Commun.* **24**(1):63 (2003).

19. Webster, D. C., *J. Coat. Technol.* **2**(15):24 (2005).
20. Hoogenboom, R. and Schubert, U. S., High-throughput synthesis equipment applied to polymer research, *Rev. Sci. Instrum.* **76**(6):062202 (2005).
21. Zhang, H., Hoogenboom, R., Meier, M. A., and Schubert, U. S., Combinatorial and high-throughput approaches in polymer science, *Meas. Sci. Technol.* **16**(1):203 (2005).
22. Cabral, J. T., Hudson, S., Harrison, C., and Douglas, J. F., Frontal photopolymerization for microfluidic applications, *Langmuir* **20**(23):10020 (2004).
23. Wu, T., Mei, Y., Cabral, J. T., Xu, C., and Beers, K. L., A new synthetic method for controlled polymerization using a microfluidic system, *J. Am. Chem. Soc.* **126**(32):9880 (2004).
24. Cygan, Z. T., Cabral, J. T., Beers, K. L., and Amis, E., Microfluidic platform for the generation of organic-phase microreactors, *Langmuir* **21**(8):3629 (2005).
25. Genzer, J., Templating surfaces with gradient assemblies, *J. Adhesion* **81**(3–4):417 (2005).
26. Eidelman, N., Raghavan, D., Forster, A. M., Amis, E. J., and Karim, A., Combinatorial approach to characterizing epoxy curing, *Macromol. Rapid Commun.* **25**:259 (2004).
27. Sormana, J. L. and Meredith, J. C., High throughput discovery of structure-property relationships for segmented polyurethanes, *Macromolecules* **37**(6): 2186 (2004).
28. Forster, A. M., Zhang, W., Crosby, A. J., and Stafford, C. M., A multi-lens measurement platform for high throughput adhesion measurement, *Meas. Sci. Technol.* **16**(1):81 (2005).
29. Chinche, A., Zhang, W., Stafford, C. M., and Karim, A., A new design for high throughput peel tests, *Meas. Sci. Technol.* **16**(1):183 (2005).
30. Takeuchi, I., Lauterbach, J., and Fasolka, M. J., Combinatorial material synthesis, *Mater. Today* **8**(10):18 (2005).
31. Konnur, R., Kargupta, K., and Sharma, A., Instability and morphology of thin liquid films on chemically heterogeneous substrates, *Phys. Rev. Lett.* **84**(5):931 (2000).
32. Reiter, G., Dewetting of thin polymer films, *Phys. Rev. Lett.*, **68**:75 (1992).
33. Brochard-Wyart, F. and Daillant, J., Drying of solids wetted by thin liquid films, *Can J. Phys.* **68**:1084 (1990).
34. de Gennes, P. G., Wetting: Statics and dynamics, *Rev. Mod. Phys.* **57**:827 (1985).
35. Hoogenboom, R., Meier, M. A. R., and Schubert, U. S., High throughput polymer screening: Exploiting combinatorial chemistry and data mining tools in catalysts and polymer development, *Macromol. Rapid Commun.* **24**:47 (2003).
36. Potyrailo, R. A., Sensors in combinatorial polymer research, *Macromol. Rapid Commun.* **25**:77 (2004).
37. Karim, A. and Kumar, S. (eds.), *Polymer Surfaces, Interfaces and Thin Films*, World Scientific, Singapore, 2000.
38. Garbassi, F., Mora, M., and Occhiello, E. (eds.), *Polymer Surfaces, from physics to technology*, Wiley, Chichester, UK, 1998.

39. Reiter, G., Unstable thin polymer films: Rupture and dewetting processes, *Langmuir* **9**:1344 (1993).
40. Zhao, W., Rafailovich, M. H., Sokolov, J., Fetters, L. J., Plano, R., Sanyal, M. K., Sinha, S. K., and Sauer, B. B., Wetting properties of thin, liquid poly(ethylene propylene) films, *Phys. Rev. Lett.* **70**:1453 (1993).
41. Ashley, K. M., Seghal, A., Amis, E., Raghavan, D., and Karim, A., Combinatorial mapping of polymer film on gradient energy surfaces, *MRS Proc. Combinatorial and Artificial Intelligence Methods in Material Science*, Boston, vol. 700, 2002, pp. 151–156.
42. Ashley, K., Meredith, J. C., Raghavan, D., Amis, E., and Karim, A., Combinatorial measurement of dewetting of polystyrene thin films, *Polym. Commun.* **44**:769 (2003).
43. Sato, Y., Study of HF-treated heavily doped Si surface using contact angle, *Jpn. J. Appl. Phys.* **33**:6508 (1994).
44. Roberson, S. V., Fahey, A. J., Sehgal, A., and Karim, A., Time of flight secondary ion mass spectrometry for high throughput characterization of biosurfaces, *Appl. Surf. Sci.* **200**:150 (2002).
45. Liedberg, B. and Tengvall, P., Molecular gradients of omega substituted alkanethiols on gold: Preparation and characterization, *Langmuir* **11**:3821 (1995).
46. Genzer, J. and Kramer, E. J., Pretransitional thining of a polymer wetting layer, *Europhys. Lett.* **44**:180 (1998).
47. Owens, D. K. and Wendt, R. C., Estimation of the surface free energy of polymers, *J. Appl. Polym. Sci.* **13**:1741 (1969).
48. Wu, S., *Polymer Interfaces and Adhesion*, Marcel Dekker, New York, 1982, p. 151.
49. Meredith, J. C., Karim, A., and Amis, E. J., Combinatorial methods for investigations in polymer material science, *MRS Bull.* **27**(4):330 (2002).
50. Xie, R., Karim, A., Douglas, J. F., Han, C. C., and Weiss, R. A., Spinodal dewetting of thin polymer films, *Phys. Rev. Lett.* **81**(6):1251 (1998).
51. Ashley, K. M., Raghavan, D., Douglas, J. F., and Karim, A., Wetting-dewetting transition line in thin polymer films, *Langmuir* **21**(21):9518 (2005).
52. Karim, A., Ashley, K. M., Douglas, J. F., and Raghavan, D., Mapping wetting-dewetting transition line in ultrathin polystyrene films combinatorially, *Polym. Mater. Sci. Eng.* **93**:900 (2005).
53. Erwim, B. M., Colby, R. H., Kamath, S. Y., and Kumar, S. K., Enhanced cooperativity between the caging temperature of glass-forming fluids, *Europhys. Lett.* **92**:185705 (2004).
54. Brochard-Wyart, F. and Daillant, J., Drying of solids wetted by thin liquid films, *Can. J. Phys.* **68**:1084 (1990).
55. Quere, D., Megilo, J. M. D., and Brochard-Wyart, F., *Science* **249**:1256 (1990).
56. Israelachvili, J., *Intermolecular and Surface Forces*, 2nd ed., Academic Press, London, 1992.
57. Mahanty, J. and Ninham, B. W., *Dispersion Forces*, Academic Press, London, 1976.
58. Tidswell, I., Rabedeau, T., Pershan, P., and Kosowksy, S., Complete wetting of a rough surface: An X-ray study, *Phys. Rev. Lett.* **66**:2108 (1991).

59. Soo, Y.-S., Koga, T., Sokolov, J., Rafailovich, M., Tolan, M., and Sinha, S., Deviation from liquidlike behavior in molten polymer films at interfaces, *Phys. Rev. Lett.* **94**:15782 (2005).

60. Choi, S. and Newby, B. Z., Alternative method for determining surface energy by utilizing polymer thin film dewetting, *Langmuir* **19**:1419 (2003).

61. Dee, G. and Sauer, B., The surface tension of polymer liquids, *Adv. Phys.* **47**:161 (1998).

62. Shin, K., Pu, Y., Rafailovich, M. H., Sokolov, J., Seeck, O. H., Sinha, S. K., Tolan, M., and Kolb, R., Correlated surfaces of free-standing polystyrene thin films, *Macromolecules* 34:5620 (2001).

63. Soles, C., Douglas, J., Jones, R., and Wu, W., Unusual expansion and concentration in ultra-thin glassy polycarbonate films, *Macromolecules* **37**:2901 (2004). (See list of references in this article regarding earlier efforts to characterize the influence of film thickness on the T_g in thin polymer films.)

64. Fryer, D., Peters, R., Kim, E., Tomaszewski, J., de Pablo, J., Nealey, P., White, C., and Wu, W., Dependence of the glass transition temperature of polymer films on interfacial energy and thickness, *Macromolecules* **34**:5627 (2001).

65. Ashley, K. M., Raghavan, D., Seghal, A., Douglas, J. F., and Karim, A. (unpublished data).

CHAPTER 9

Combinatorial Materials Science: Challenges and Outlook

BALAJI NARASIMHAN and SURYA K. MALLAPRAGADA

Institute for Combinatorial Discovery
Department of Chemical and Biological Engineering
Iowa State University
Ames, IA

MARC D. PORTER

Department of Chemistry and Biochemistry
Center for Combinatorial Science at The Biodesign Institute
Arizona State University
Tempe, Arizona

9.1. OVERVIEW

As exemplified by the topics detailed in the preceding chapters, materials science has made enormous contributions to the technological revolution of the last century and will lead the next breakthroughs in areas central to energy, healthcare, transportation, food safety and security, and antiterrorism. Futuristic innovations, however, will demand new classes of materials with improved or even unforeseen properties and levels of performance in order to function as the next generation of highly active, durable catalysts, nanomachines, molecularly engineered surfaces, and medicine. Accelerating these developments will nevertheless require innovations in materials design, discovery, and analysis.

The discovery of new materials traditionally has relied on a "one experiment at a time" approach in which a small collection of materials are synthesized and carefully evaluated in order to make incremental improvements in properties or performance. Occasionally, serendipity leads to an innovative transition

Combinatorial Materials Science, Edited by Balaji Narasimhan, Surya K. Mallapragada, and Marc D. Porter

in these efforts by revealing material constructs that lie beyond convention. Combinatorial science (CombiSci) is a disruptive approach that enables a multitude of materials and the impact of varied preparative conditions to be evaluated in a single experiment. CombiSci therefore embodies the use of massively parallel strategies for the creation and high-throughput testing of enormous numbers of samples (i.e., libraries) for accelerated discovery. Combinatorial techniques are invaluable for generating potential solutions to complex problems possessing a vast search space, and represent a paradigm shift from the laborious, one-sample-at-a-time approaches [1–6]. Experiments that previously required months or years to accomplish can now be performed in days to weeks. Although initially devised to accelerate drug discovery, combinatorial science is gaining a central position in materials research and has the potential to irreversibly alter and enhance materials design and discovery.

9.2. OUTLOOK

CombiSci is emerging as a vital pathway to unravel the complexities of fundamental structure–function relationships in high-performance materials. Chapters 2–8 covered a wide range of material issues and the new tools being devised for the high-throughput investigation of materials properties, highlighting the significance of CombiSci methodology. Its importance is further underscored by the billions of dollars in investments by industry leaders [2,4]. For example, in 1998, Dow and Symyx partnered in a $120 million collaborative program to develop combinatorial catalysis. The vitality of this arena can also be gauged by an examination of the chemical instruments market ($47 billion in 1998), which includes the hardware for both library design and screening [7]. While projected to grow annually at 3–5% in the next 5–10 years, the predicted expansion of the subsector for CombiSci instruments is 12–20% per year. Another insightful measure is the meteoric rise of the *Journal of Combinatorial Chemistry* (JCC), launched in 1999. The *2004 ISI Journal Citation Report* indicates that of the 125 multidisciplinary chemistry journals, JCC already is ranked 11th in impact factor and is second among applied chemistry journals. Several other journals have recognized the importance of CombiSci and have devoted special issues to this area. Among these are *Macromolecular Rapid Communications* in 2003 and 2004 [Vol. **24**(1) in 2003 and Vol. **25**(1) in 2004] and *MRS Bulletin* [Vol. **27**(4) in 2002]. CombiSci has also begun to have a strong presence at scientific conferences and meetings. A few recent examples include the 227th, 229th, and 231st ACS meetings in 2004–2006, which featured symposia on combinatorial approaches to materials; Gordon research conferences in 2004 and 2005 focused on combinatorial and high-throughput materials science, and MRS symposia in 2001 and 2003 focused on the area of combinatorial and artificial intelligence methods in materials science. These metrics are forceful testimonies of the far-reaching impact of CombiSci research.

The chapters in this book have focused on two themes: (1) development and exploitation of CombiSci as a process for systematic and accelerated investigation of new phenomena and of the complex structure–function interplay in materials and (2) creation of new library design strategies for materials processing and of high-throughput tools for rapid screening. These treatments have emphasized innovations in catalysts, biomaterials, and nanomaterials, as well as informatics approaches to analyze and mine CombiSci data.

By far, the best-known application of CombiSci has been in the pharmaceutical industry, the first market sector to use CombiSci in its search for new drugs. While solution processes test drug candidates, materials characterization involves determination of properties like modulus or interfacial reactivity. As a consequence, combinatorial materials science demands screening tools well beyond those in the drug discovery arsenal, and we envision that such developments will emerge as an increasing active research endeavor for the next several decades.

9.3. CHALLENGES

9.3.1. Technology

CombiSci is an inherently technology-driven enterprise, requiring advanced robotics and parallel synthesis tools for library fabrication and a plethora of rapid and high-throughput screening methods for materials analysis. However, a major technological breakthrough is necessary to elucidate the atom-level structural and compositional complexities of large-scale materials libraries in the search for nextgeneration materials. We believe that the family of scanning probe microscopies will play a key role in this area. Moreover, recent breakthroughs in the atom probe microscope (APM) will provide a new tool with which to perform atom-scale characterization of chemistry and structure of materials at an unprecedented level of detail—indeed, the first market entrants of this technology. It is important to note, however, that advancements in sample preparation are critical to extend APM to combinatorial investigations of materials chemistry and structure [8]. Other challenges include the characterization of samples with low conductivity (e.g., organic and biological materials) and the three-dimensional spatial reconstruction and visualization atom probe data. While existing approaches enable viewing of only a small number of ions (~30,000), high-throughput techniques will likely require the ability to rapidly project more than 1 million ion positions. Another challenge is that the algorithms used in smoothing data to accommodate the gaps in the sequential evolution of the tip and the resultant gaps in the reconstruction are very time-consuming and hinder analysis and interpretation. As these challenges are overcome, we firmly believe that APM will become a central component in the toolbox for combinatorial materials science.

9.3.2. Informatics

The sheer quantity and complexity of data generated from CombiSci experiments leads to a data analysis bottleneck. The radical changes in information generation and structure driven by CombiSci require sophisticated informatics tools to digest massive datasets and advanced statistical analysis methods to address multidimensional error analysis and experimental design. These needs will drive close collaborations between materials and computer scientists, database personnel, and statisticians for the development of highly effective tools that will not only support and integrate computational materials science (i.e., informatics), enhance the structured design of experiments (i.e., combinatorial experimentation), but also facilitate their access the materials science community.

9.3.3. Education

While combinatorial science has grown as a potent field that blurs the traditional boundaries of chemistry, biology, materials science, and engineering, the employment demands of industry have outpaced the growth of the talent pool. This shortage has given birth to a number of workshops, offered by leading conferences (e.g., PittCon, CombiChem Europe), and training groups that are formulated to fill the deficiency. At present, however, only a few academic institutions have formally organized the rigorous research and education programs dictated by CombiSci, and almost all are focused on drug discovery. A clear need therefore exists to (1) broadly integrate the scientific fundamentals and interdisciplinary underpinnings required to develop and apply CombiSci concepts to areas of National Need in materials and (2) proactively respond to the growing demand for a highly skilled workforce. In this context, new courses and laboratories must be designed to expose students to the many facets of this area. Educators in biology, chemistry, physics, mathematics, computer science, and engineering will need to work together to develop these courses and design appropriate curricula in order to train students from diverse backgrounds, emphasizing a balanced graduate-level education as well as strong scientific grounding.

9.4. SUMMARY

In summary, combinatorial science is a highly potent methodology that will result in new generations of materials. CombiSci enables researchers to move between the gap from the realm of "known knowns" to that of "unknown unknowns," thereby adding a new dimension to conventional experimental and modeling approaches. If successful, "game changing" discoveries will result and materials with new and potentially unforeseen properties and levels of performance will be realized.

REFERENCES

1. Szostak, J., Combinatorial chemistry: Special thematic issue. *Chem. Rev.* **97**:347–509 (1997).
2. Borman, S., Combinatorial chemistry, *Chem. Eng. News* **76**:47–54 (1998).
3. Watkins, K., Strength in numbers, *Chem. Eng. News* **80**:30–34 (2001).
4. Hewes, J. D., Herring, L., Schen, M. A., Cuthill, B., and Sienkiewicz, R., *Combinatorial Discovery of Catalysts: An ATP Position Paper Developed from Industry Input*, Gaithersburg, MD, 1998.
5. Cawse, J. N., Experimental strategies for combinatorial and high-throughput materials development, *Acc. Chem. Res.* **34**:213–221 (2001).
6. Braeckmans, K., Smedt, S. C., Leblans, M., Roelant, C., Pauwels, R., and Demeester, J., Scanning the code, *Modern Drug Discov.* **6**:28–32 (2003).
7. U.S. Department of Commerce, *1996 Annual Survey of Manufacturers*, Washington, DC, 1998.
8. Jacoby, M., Atomic imaging turns 50, *Chem. Eng. News* **83**(48):13–16 (2005).

INDEX

Combinatorial Materials Science, Edited by Balaji Narasimhan, Surya K. Mallapragada, and Marc D. Porter

Printed and bound by CPI Group (UK) Ltd, Croydon, CR0 4YY
16/09/2026
14963443-0001